Studies in Computational Intelligence

Volume 619

Series editor

Janusz Kacprzyk, Polish Academy of Sciences, Warsaw, Poland
e-mail: kacprzyk@ibspan.waw.pl

About this Series

The series "Studies in Computational Intelligence" (SCI) publishes new developments and advances in the various areas of computational intelligence—quickly and with a high quality. The intent is to cover the theory, applications, and design methods of computational intelligence, as embedded in the fields of engineering, computer science, physics and life sciences, as well as the methodologies behind them. The series contains monographs, lecture notes and edited volumes in computational intelligence spanning the areas of neural networks, connectionist systems, genetic algorithms, evolutionary computation, artificial intelligence, cellular automata, self-organizing systems, soft computing, fuzzy systems, and hybrid intelligent systems. Of particular value to both the contributors and the readership are the short publication timeframe and the worldwide distribution, which enable both wide and rapid dissemination of research output.

More information about this series at http://www.springer.com/series/7092

Roger Lee
Editor

Applied Computing & Information Technology

Editor
Roger Lee
Software Engineering and Information Institute
Central Michigan University
Mount Pleasant, MI
USA

ISSN 1860-949X ISSN 1860-9503 (electronic)
Studies in Computational Intelligence
ISBN 978-3-319-26394-6 ISBN 978-3-319-26396-0 (eBook)
DOI 10.1007/978-3-319-26396-0

Library of Congress Control Number: 2015955879

Springer Cham Heidelberg New York Dordrecht London

Printed on acid-free paper

Springer International Publishing AG Switzerland is part of Springer Science+Business Media (www.springer.com)

Foreword

The purpose of the 3rd ACIS International Conference on Applied Computing and Information Technology (ACIT 2015) held on July 12–16, 2015, in Okayama, Japan, was to bring together researchers, scientists, engineers, industry practitioners, and students to discuss, encourage, and exchange new ideas, research results, and experiences on all aspects of Applied Computers & Information Technology, and to discuss the practical challenges encountered along the way and the solutions adopted to solve them. The conference organizers have selected the best 14 papers from those papers accepted for presentation at the conference in order to publish them in this volume. The papers were chosen based on the review scores submitted by members of the program committee and underwent further rigorous rounds of review.

In chapter "Improving Relevancy Filter Methods for Cross-Project Defect Prediction," Kazuya Kawata, Sousuke Amasaki, and Tomoyuki Yokogawa propose and examine a new relevancy filter method using an advanced clustering method density-based spatial clustering (DBSCAN). The results suggested that exploring advanced clustering algorithms could contribute to cross-protect defect predictions.

In chapter "A Game Framework Supporting Automatic Functional Testing for Games," Woei-Kae Chen, Chien-Hung Liu, Ping-Hung Chen, and Chia-Sheng Hsu propose a method that automates game testing. Since games are usually built on top of game frameworks, the idea is to enhance a game framework with a testing layer, which can execute (playback) test scripts that perform user events and assert the correctness of the game. The results showed that when repeated testings are necessary, an automatic testing (either writing or recording test scripts) can reduce human cost.

In chapter "Feature Extraction and Cluster Analysis Using N-gram Statistics for DAIHINMIN Programs," Seiya Okubo, Takaaki Ayabe, and Tetsuro Nishino elucidate the characteristics of the computer program to play daihinmin, which is a popular Japanese card game. They propose a method to extract feature values using the n-gram statistics and a cluster analysis method using the feature values. They show that our proposed method can successfully cluster daihinmin programs with high probability.

In chapter "Prediction Interval of Cumulative Number of Software Faults Using Multilayer Perceptron," Begum Momotaz and Tadashi Dohi focus on a prediction problem with the common multilayer perceptron neural networks and derive the predictive interval of the cumulative number of software faults in sequential software testing. They apply the well-known back-propagation algorithm for feed-forward neural network architectures and the delta method to construct the prediction intervals.

In chapter "Heuristics for Daihinmin and Their Effectiveness," Yasuhiro Tajima and Kouzou Tagashira show four heuristics for the card game "daihinmin" and evaluate their effectiveness by some experiments. The result is that their heuristics are effective and our program is as strong as Monte Carlo algorithm.

In chapter "Cluster Analysis for Commonalities Between Words of Different Languages," Jennifer Daniels, Doug Nye, and Gongzhu Hu present a clustering analysis of $n = 815$ commonly used words in eight different languages including five Western languages (English, German, French, Spanish, and Italian), an ancient language (Latin), an Asian language (Japanese), and the language of a Native American tribe (Ojibwa). The results tend to support one of the leading theories describing how Native American tribes, specifically the Ojibwa people, arrived in North America.

In chapter "The Design and Partial Implementation of the Dementia-Aid Monitoring System Based on Sensor Network and Cloud Computing Platform," Dingkun Li, Hyun Woo Park, Minghao Piao, and Keun Ho Ryu aim to design a comprehensive, unobtrusive, real-time, low-cost but effective monitoring system to help caregivers for their daily healthcare work. The system design has been finished and the entire experimental environment including sensor network and cloud model has been set up in their laboratory to collect simulated data and one healthcare center to collect real data.

In chapter "Experimental Use of Learning Environment by Posing Problem for Learning Disability," Sho Yamamoto, Tsukasa Hirashima, and Akio Ogihara describe about a design of learning environment based on the information structure and a realization of problem posing for learning disability. They designed and developed software by modeling information structure of subject that is operated on software. In this research, they aimed at the domain of education and developed a learning environment for posing arithmetic word problem.

In chapter "An Approach to Estimating Decision Complexity for Better Understanding Playing Patterns of Masters," Akira Takeuchi, Masashi Unoki, and Hiroyuki Iida propose a method for estimating decision complexity of positions by using correlation coefficient between two evaluation values of the root node and leaf level of a game tree. The results show that the proposed idea is a promising way to better understand playing patterns of masters, for example, identifying the right moment for possibility of changing strategies such as speculative play and early resignation.

In chapter "A Routing Algorithm for Distributed Key-Value Store Based on Order Preserving Linear Hashing and Skip Graph," Ken Higuchi, Makoto Yoshida, Makoto Yoshida, and Tatsuo Tsuji propose a routing algorithm for the distributed key-value store based on the order preserving linear hashing and Skip graph. By using the proposed algorithm, reduction of the number of hops are expected. From the experimental results, it proved that the proposed algorithm is effective.

In chapter "Effects of Lower Limb Cooling on the Work Performance and Physiological Responses during Maximal Endurance Exercise in Humans," Keiko Inoue, Masashi Kume, and Tetsuya Yoshida investigate the effects of lower limb cooling on the work performance and physiological responses during maximal endurance exercise in humans. Eight male subjects underwent a maximal aerobic test using graded exercise on a cycle ergometer.

In chapter "Demonstration of Pedestrian Movement Support System Using Visible Light Communication," Saeko Oshiba, Shunsuke Iki, Hirotoshi Kii, Hiroki Watanabe, Yusuke Murata,Yuki Nagai, Jun Yamazaki, Masato Yoshihisa, Yoji Kitani, Saori Kitaguchi, and Kazunari Morimoto construct a pedestrian support system for the visually impaired that uses visible light communication (VLC) from self-illuminating bollards. They created a prototype of the navigation system using VLC. The prototype was evaluated by visually impaired users, clarifying the effectiveness of the system.

In chapter "The Image Sharpness Metric via Gaussian Mixture Modeling of the Quaternion Wavelet Transform Phase Coefficients with Applications," Yipeng Liu and Weiwei Du make use of Gaussian mixture model (GMM) to describe the coefficients distribution of the quaternion wavelet transform (QWT). Derived from the parameters in GMM, the metric is proposed to find the relationship between the image blur degree and the distribution histograms of high-frequency coefficients.

In chapter "Generating High Brightness Video Using Intermittent Illuminations for Dark Area Surveying," Yoshiyama Hitomi, Iwai Daisuke, and Sato Kosuke propose a method for reducing power consumption of necessary illuminations for a camera on a remote control robot. Experiments to confirm operation of the proposed method show that an image sequence as if illuminations are always lighting can be generated in the proposed method.

It is our sincere hope that this volume provides stimulation and inspiration and that it will be used as a foundation for works to come.

July 2015

Takaaki Goto
Ryutsu Keizai University, Japan

Contents

Contributors

Sousuke Amasaki Okayama Prefectural University, Okayaka, Japan

Takaaki Ayabe Graduate School of Electro-Communications, University of Electro-Communications, Chofu, Tokyo, Japan

Ping-Hung Chen Department of Computer Science and Information Engineering, National Taipei University of Technology, Taipei, Taiwan, Republic of China

Woei-Kae Chen Department of Computer Science and Information Engineering, National Taipei University of Technology, Taipei, Taiwan, Republic of China

Jennifer Daniels Department of Computer Science, Central Michigan University, Mount Pleasant, MI, USA

Tadashi Dohi Department of Information Engineering, Graduate School of Engineering Hiroshima University, Higashi-Hiroshima, Japan

Weiwei Du Department of Information Science, Kyoto Institute of Technology, Kyoto, Japan

Ken Higuchi Graduate School of Engineering, University of Fukui, Fukui, Japan

Tsukasa Hirashima Graduate School of Engineering, Hiroshima University, Higashi-Hiroshima City, Hiroshima, Japan

Chia-Sheng Hsu Department of Computer Science and Information Engineering, National Taipei University of Technology, Taipei, Taiwan, Republic of China

Gongzhu Hu Department of Computer Science, Central Michigan University, Mount Pleasant, MI, USA

Hiroyuki Iida School of Information Science, Japan Advance Institute of Science and Technology, Ishikawa, Japan

Shunsuke Iki Kyoto Institute of Technology, Sakyoku, Kyoto, Japan

Keiko Inoue Kyoto Institute of Technology, Matsugasaki, Sakyo-Ku, Kyoto, Japan

Daisuke Iwai Graduate School of Engineering Science, Osaka University, Toyonaka, Osaka, Japan

Kazuya Kawata Okayama Prefectural University, Okayaka, Japan

Hirotoshi Kii Kyoto Institute of Technology, Sakyoku, Kyoto, Japan

Saori Kitaguchi Kyoto Institute of Technology, Sakyoku, Kyoto, Japan

Yoji Kitani Kyoto Institute of Technology, Sakyoku, Kyoto, Japan

Masashi Kume Kyoto Bunkyo Junior College, Makishima, Senzoku80, Uji, Kyoto, Japan

Dingkun Li Database/Bioinformatics Lab, School of Electrical and Computer Engineering, Chungbuk National University, Cheongju, South Korea

Chien-Hung Liu Department of Computer Science and Information Engineering, National Taipei University of Technology, Taipei, Taiwan, Republic of China

Yipeng Liu College of Information Engineering, Zhejiang University of Technology, Hangzhou, China

Naoyuki Miyamoto Graduate School of Engineering, University of Fukui, Fukui, Japan

Begum Momotaz Department of Information Engineering, Graduate School of Engineering Hiroshima University, Higashi-Hiroshima, Japan

Kazunari Morimoto Kyoto Institute of Technology, Sakyoku, Kyoto, Japan

Yusuke Murata Kyoto Institute of Technology, Sakyoku, Kyoto, Japan

Yuki Nagai Kyoto Institute of Technology, Sakyoku, Kyoto, Japan

Tetsuro Nishino Graduate School of Informatics and Engineering, University of Electro-Communications, Chofu, Tokyo, Japan

Doug Nye Department of Computer Science, Central Michigan University, Mount Pleasant, MI, USA

Akio Ogihara Department of Informatics, Faculty of Engineering, Kinki University, Higashi-Hiroshima City, Hiroshima, Japan

Seiya Okubo School of Management and Information, University of Shizuoka, Shizuoka, Japan

Saeko Oshiba Kyoto Institute of Technology, Sakyoku, Kyoto, Japan

Hyun Woo Park Database/Bioinformatics Lab, School of Electrical and Computer Engineering, Chungbuk National University, Cheongju, South Korea

Minghao Piao Department of Computer Engineering, Dongguk University Gyeongju Campus, Gyeongju, South Korea

Keun Ho Ryu Database/Bioinformatics Lab, School of Electrical and Computer Engineering, Chungbuk National University, Cheongju, South Korea

Kosuke Sato Graduate School of Engineering Science, Osaka University, Toyonaka, Osaka, Japan

Kouzou Tagashira Graduate School of Computer Science and Systems Engineering, Okayama Prefectural University, Okayama, Japan

Yasuhiro Tajima Department of Systems Engineering, Okayama Prefectural University, Okayama, Japan

Akira Takeuchi School of Information Science, Japan Advance Institute of Science and Technology, Ishikawa, Japan

Tatsuo Tsuji Graduate School of Engineering, University of Fukui, Fukui, Japan

Masashi Unoki School of Information Science, Japan Advance Institute of Science and Technology, Ishikawa, Japan

Hiroki Watanabe Kyoto Institute of Technology, Sakyoku, Kyoto, Japan

Sho Yamamoto Department of Informatics, Faculty of Engineering, Kinki University, Higashi-Hiroshima City, Hiroshima, Japan

Jun Yamazaki Kyoto Institute of Technology, Sakyoku, Kyoto, Japan

Tomoyuki Yokogawa Okayama Prefectural University, Okayaka, Japan

Makoto Yoshida Graduate School of Engineering, University of Fukui, Fukui, Japan

Tetsuya Yoshida Kyoto Institute of Technology, Matsugasaki, Sakyo-Ku, Kyoto, Japan

Masato Yoshihisa Kyoto Institute of Technology, Sakyoku, Kyoto, Japan

Hitomi Yoshiyama Graduate School of Engineering Science, Osaka University, Toyonaka, Osaka, Japan

Improving Relevancy Filter Methods for Cross-Project Defect Prediction

Kazuya Kawata, Sousuke Amasaki and Tomoyuki Yokogawa

Abstract Context: Cross-project defect prediction (CPDP) research has been popular. One of the techniques for CPDP is *a relevancy filter* which utilizes clustering algorithms to select a useful subset of the cross-project data. Their performance heavily relies on the quality of clustering, and using an advanced clustering algorithm instead of simple ones used in the past studies can contribute to the performance improvement. Objective: To propose and examine a new relevancy filter method using an advanced clustering method DBSCAN (Density-Based Spatial Clustering). Method: We conducted an experiment that examined the predictive performance of the proposed method. The experiments compared three relevancy filter methods, namely, Burak-filter, Peters-filter, and the proposed method with 56 project data and four prediction models. Results: The predictive performance measures supported the proposed method. It was better than Burak-filter and Peters-filter in terms of AUC and g-measure. Conclusion: The proposed method achieved better prediction than the conventional methods. The results suggested that exploring advanced clustering algorithms could contribute to cross-project defect prediction.

Keywords Cross-project defect prediction · Relevancy filter · DBSCAN

K. Kawata · S. Amasaki (✉) · T. Yokogawa
Okayama Prefectural University, 111 Kuboki, Soja,
Okayaka 719-1197, Japan
e-mail: amasaki@cse.oka-pu.ac.jp

K. Kawata
e-mail: cd26018m@cse.oka-pu.ac.jp

T. Yokogawa
e-mail: t-yokoga@cse.oka-pu.ac.jp

R. Lee (ed.), *Applied Computing & Information Technology*,
Studies in Computational Intelligence 619,
DOI 10.1007/978-3-319-26396-0_1

1 Introduction

Software systems become larger and larger in recent years. A large software system consists of many modules, and some of the modules might have potential defects. Thus, it is important to detect and remove the defects by carefully testing and reviewing the modules as early as possible. We can perform the testing and reviewing activities more efficiently by spotting suspicious modules—it is referred to as "software defect prediction".

Software defect prediction has been one of the popular research topics in software engineering area. Most of the existing prediction methods leverage supervised learning techniques that require the project data including the defects and the process/product metrics of modules. That is to say, these methods are applicable to only the project which can provide its project data. However, there are also many projects whose project data is unavailable; for example, a new project cannot provide the data of defects and process metrics. Therefore, there have been some studies to overcome the lack of the data by using the data from other projects (*cross-project data*)—they are called the cross-project defect prediction (CPDP).

Since different projects have different size and different distributions of dataset, a dataset from a project may not work for the defect prediction at another project. Therefore, many CPDP methods perform a data processing to a cross-project data for enhancing the defect prediction.

One of the useful data processing techniques for CPDP is the relevancy filter, which focuses on a part of the cross-project data. Turhan et al. [9] used K-nearest neighbors (K-NN) as a data processing for CPDP (they called Burak-filter): for each module of a target project, they select similar modules from a cross-project data by using K-NN and predict the defect proneness of the module through supervised learning techniques with the data of the selected modules. Peters et al. [7] proposed Peters-filter, which mixes K-means clustering method with K-NN to improve the predictive performance of [9].

While they could contribute to better prediction, the predictive performance has still been challenged. Their performance heavily relies on the quality of subset selection, and using an advanced clustering method instead of simple ones like K-means can contribute to the performance improvement. However, an application of other algorithms on the relevancy filter has not been studied. This study thus proposed and examined a new relevancy filter method based on one of the advanced clustering algorithms, DBSCAN (the Density-Based Spatial Clustering).

The remainder of this paper is organized as follows: Sect. 2 describes the existing relevancy filter methods and the method based on DBSCAN. Section 3 gives experiment settings for performance evaluations of the relevancy filter methods. Section 4 presents the results and the discussion. Section 5 describes the related work. Finally, Sect. 6 concludes this paper.

2 Cross-Company Defect Prediction Methods

This section introduces three relevancy filter methods. Burak-filter and Peters-filter were proposed in the past studies. Our proposed method was based on a advanced clustering method DBSCAN.

2.1 *Burak-Filter*

Burak-filter was proposed by Turhan et al. [9]. Although the authors connected it with Naive Bayes Classifier, the filter method was independent of prediction models. The procedure of Burak-filter is as follows:

1. Find *k* neighbors from the cross-project data for each record in the target project data based on Euclidean distance, and
2. Collect the selected neighbors without duplication into a new cross-project data.

2.2 *Peters-Filter*

Peters-filter was proposed by Peters et al. [7]. It utilized K-means clustering method to group records of the target and cross-project data for finding similar records to ones in the target project data. The procedure of Peters-filter is as follows:

1. Combine the target project data and the cross-project data,
2. Divide the combined-project data into sub-clusters by using K-means clustering method based on Euclidean distance,
3. Select sub-clusters which have at least one record of the target project data,
4. From each selected sub-cluster,
   (a) Find the nearest neighbor of each record of the cross-project data from the target project data in the same cluster,
   (b) Select the nearest neighbor of each record of the selected target project data from the cross-project data which selects that target record as its nearest neighbor, and
5. Collect the selected neighbors from the sub-clusters into a new cross-project data.

Peters-filter does not necessarily find neighbors for all records of the target project data while Burak-filter does. This treatment filters out non-representative records in a target project data and improves the appropriateness of selected records.

2.3 A Filter Method Based on DBSCAN

DBSCAN (Density-Based Spatial Clustering) [1] is one of the advanced clustering algorithms. It can remove noisy instances based on the density of instances as same as Peters-filter. It was thus expected that at least it can find better clusters as good as Peters-filter could. Therefore, we adopted DBSCAN for a relevancy filter method.

DBSCAN finds sub-clusters based on the density of records on a vector space such as Euclidean space. The records in a high-density area are called core sample. A cluster consists of the core samples close to each other. Here, DBSCAN defines the high density with two parameters: the distance which determines whether two records are close to and the number of records which determines whether a core sample is in a dense area. DBSCAN considered the record as outlier which is not a core sample and far from any core sample. This noise reduction may help filtering out non-representative records.

The procedure we applied DBSCAN is as follows:

1. Combine the target project data and the cross-project data,
2. Find sub-clusters by using DBSCAN,
3. Select sub-clusters which have at least one record of the target project data, and
4. Collect the records of cross-project data in the selected sub-clusters into a new cross-project data.

This procedure was similar to Peters-filter in that the target project data and the cross-project data were clustered simultaneously. In contrast, the procedure was simple because DBSCAN itself removes irrelevant records.

3 Experiment Settings

This section describes an experiment settings we adopted. All settings followed to [7]. The one different point is the use of AUC for evaluating predictive performance.

3.1 Datasets

This study used the same datasets used in [7]. They are 56 datasets from open-source software projects. The software of those projects were written in Java, and one record corresponds to one class. All datasets were collected by Jureczko and Madeyski [4] and served on the PROMISE repository. They have 20 static code metrics measured by ckjm.[1] We divided the datasets into two groups, namely, Test datasets and Training datasets, according to [7]. Test datasets consist of 21 datasets

[1]http://www.spinelis.gr/sw/ckjm.

Table 1 Datasets for tests

Project	Sample size (# of classes)	Defect rate (%)	Project	Sample size (# of classes)	Defect rate (%)
berek	43	37	pdftranslator	33	45
ckjm	10	50	serapion	45	20
e-learning	64	8	skarbonka	45	20
forrest-0.6	6	17	sklebagd	20	60
forrest-0.7	29	17	systemdata	65	14
forrest-0.8	32	6	szybkafucha	25	56
intercafe	27	15	termoproject	42	31
kalkulator	27	22	workflow	39	51
nieruchomosci	27	37	wspomaganiepi	18	67
pbeans1	26	77	zuzel	29	45
pbeans2	51	20			

having less than 100 records. Training datasets consist of the others 35 datasets. We regard Test datasets as our target project data and Training datasets as our cross-project data. Tables 1 and 2 show basic statistics of the datasets.[2]

3.2 *Performance Measures*

We used the three performance measures: area under the ROC curve (AUC), g-measure, and F-measure. These measures take a value between 0 and 1. The higher the value means, the better performance. AUC should take more than 0.5 that implies the prediction is better than random guessing. The g-measure was a harmonic mean of recall and specificity. Specificity is a statistical measure defined as $TN/(TN + FP)$ where TN and FP mean *true negative* and *false positive*, respectively. It was used in [7]. Turhan et al. [9] also used a performance measure based on recall and specificity. F-measure is a common performance measure for classification though it was not used in [7, 9].

This study regarded AUC as the priority performance measure. Few studies adopted it though Rahman et al. [8] noted its advantages for defect prediction setting. AUC has advantages against major performance measures such as F-measure in that a threshold for binary classification and in that an imbalanced defect instance proportion.

[2]The original study reported wrong numbers for the sample size of `log4j-1.1`, `velocity-1.4`, and `velocity-1.5`. This might cause the difference of predictive performance between [7] and this study.

Table 2 Datasets for train

Project	Sample size (# of classes)	Defect rate (%)	Project	Sample size (# of classes)	Defect rate (%)
ant-1.3	125	16	poi-2.0	314	12
ant-1.4	178	22	poi-2.5	385	64
ant-1.5	293	11	poi-3.0	442	64
ant-1.6	351	26	prop-6	660	10
camel-1.0	339	4	synapse-1.0	157	10
camel-1.2	608	36	synapse-1.1	222	27
camel-1.4	872	17	synapse-1.2	256	34
ivy-1.1	111	57	velocity-1.4	196	75
ivy-1.4	241	7	velocity-1.5	214	66
jedit-3.2	272	33	velocity-1.6	229	34
jedit-4.0	306	25	xalan-2.4	723	15
jedit-4.1	312	25	xalan-2.5	803	48
jedit-4.2	367	13	xalan-2.6	885	46
log4j-1.0	135	25	xerces-1.2	440	16
log4j-1.1	109	34	xerces-1.3	453	15
lucene-2.0	195	47	xerces-1.4	588	74
lucene-2.2	247	58	xerces-init	162	48
poi-1.5	237	59			

3.3 Prediction Models

CPDP studies used several prediction models: Naive Bayes Classifier for Burak-filter [9] and Transfer Naive Bayes [5], Logistic Regression for TCA+ [6].

Peters et al. [7] recommended Random Forests among four prediction models. This study used the same prediction models as them because we also used the same datasets as them. The four prediction models are as follows:

- Logistic Regression
- Random Forests
- Naive Bayes
- k-Nearest Neighbors (with $k = 1$)

This study used `scikit-learn`[3] and Python for the experiment. We used default values for all parameters of the models.

[3]http://scikit-learn.org/.

3.4 Experimental Design

We conducted the experiment for performance comparisons following the same procedure as Peters et al. [7]. The procedure is as follows:

1. Select one project data as a target project data from Test datasets,
2. Select and combine all projects data from Train datasets,
3. Perform cross-project defect prediction with Burak-filter, Peters-filter, and the proposed method,
4. Measure predictive performance with the four prediction models, and
5. Repeat the above procedure till all project data are selected as a target project data from Test datasets.

Burak-filter has a parameter k that is the number of neighbor instances to be selected. This study used the same $k = 10$ as the past studies [7, 9] did. Peters filter has a parameter k' that is the number of clusters. This study used the same adjustment as the original study that each cluster tends to have 10 records in average. For DBSCAN, the number of minimum records was set to 10, and the distance is set to 1.0.

Peters-filter and the proposed method do not guarantee that an obtained cross-project data has at least one record with a defect. Accordingly, we first divided the cross-project data into a defective subset and a non-defective subset. Then, we applied these methods to each subset and combined the results to make a new cross-project data.

4 Results and Discussion

Figure 1 shows the predictive performance of the three methods and prediction without any filters in terms of AUC. These boxplots are based on the results from 21 target project data. The predictive performance was different among the project data, and the boxplots depict their variances. Figure 1 has four slots for four prediction models. Each slot has four box-plots; from the left, non-filter, Burak-filter, Peters-filter, and the proposed method. Table 3 shows median values of Fig. 1. It also contains precision and recall, which are base measures of F-measure.

Most of the boxes place above 0.5 and indicate that all prediction models showed moderate performance in terms of AUC. There was no clear difference among prediction models. The prediction without any filter showed the worst predictive performance for all prediction models except for K-NN. It means that the relevancy methods worked well. Peters-filter showed worse predictive performance than Burak-filter. The proposed method showed the best performance among

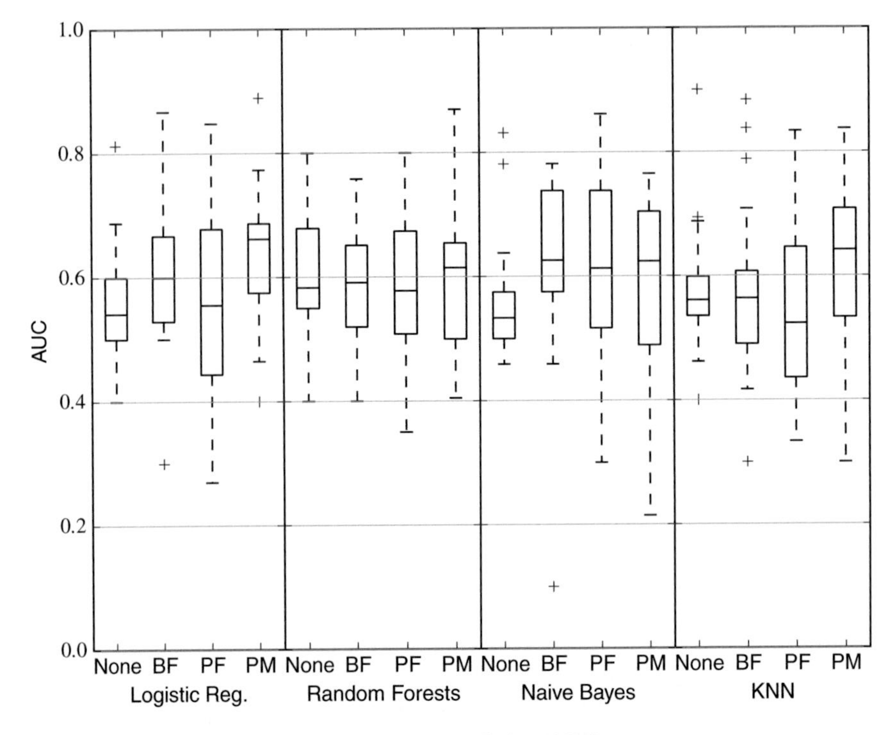

Fig. 1 Predictive performance of cross-project prediction (AUC)

Table 3 Median values of performance measures

Method	Prediction	AUC	F	Precision	Recall	g
No filter	LR	0.542	0.167	0.600	0.100	0.182
	RF	0.583	0.286	0.500	0.200	0.331
	NB	0.533	0.143	0.600	0.077	0.143
	KNN	0.561	0.316	0.300	0.333	0.472
Burak filter	LR	0.600	0.333	0.600	0.231	0.361
	RF	0.591	0.333	0.429	0.300	0.448
	NB	0.626	0.462	0.600	0.462	0.558
	KNN	0.564	0.381	0.500	0.333	0.456
Petres filter	LR	0.556	0.424	0.357	0.500	0.500
	RF	0.578	0.414	0.375	0.500	0.516
	NB	0.612	0.471	0.474	0.600	0.544
	KNN	0.524	0.385	0.400	0.500	0.494
DBSCAN method	LR	0.662	0.462	0.500	0.444	0.559
	RF	0.615	0.381	0.304	0.400	0.547
	NB	0.624	0.444	0.353	0.500	0.572
	KNN	0.641	0.424	0.400	0.600	0.597

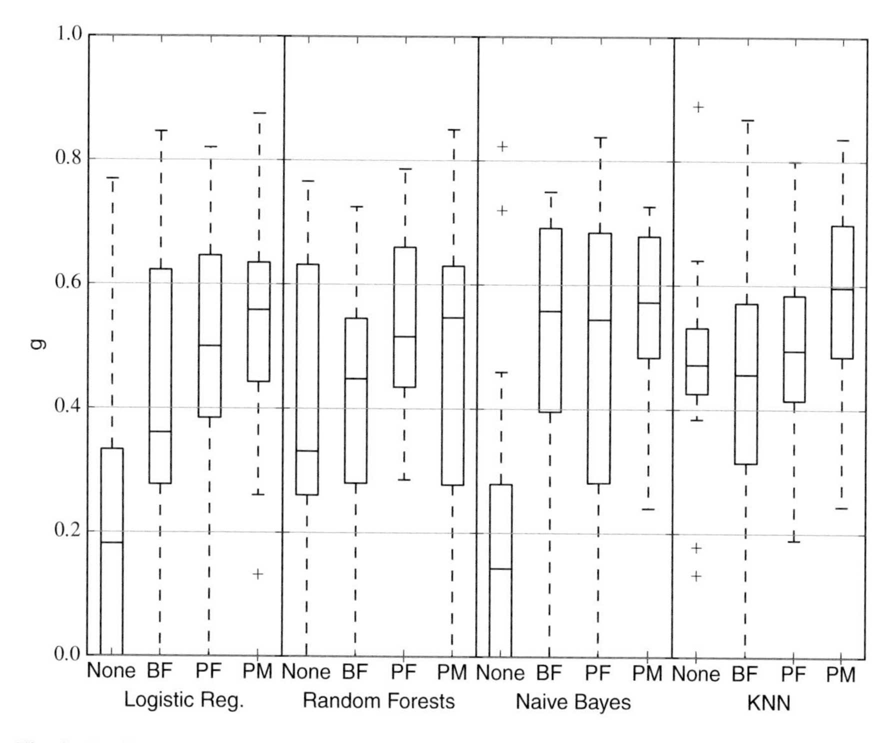

Fig. 2 Predictive performance of cross-project prediction (g-measure)

the relevancy methods. Table 3 shows the proposed method improved AUC by approximately 3–12 % points in comparison to the prediction without filtering.

Figure 2 shows g-measure. In [7], the authors compared them with g-measure and claimed that Peters-filter was better than Burak-filter. Figure 2 supported this claim while AUC supported Burak-filter. The proposed method also showed the best performance among the relevancy methods. Table 3 shows the proposed method improved g-measure by approximately 4–10 % points in comparison to Peters-filter.

Figure 3 shows F-measure. All relevancy filter methods showed better performance than prediction without any filter. The proposed method showed better performance than the others with Logistic Regression and K-NN while it showed worse performance with the other prediction models. The advantage of the proposed method was marginal.

The primary performance measure AUC supported the proposed method. The g-measure also supported the proposed method. Therefore, we concluded that the proposed method achieved better performance than the other relevancy filter methods while F-measure showed no clear difference.

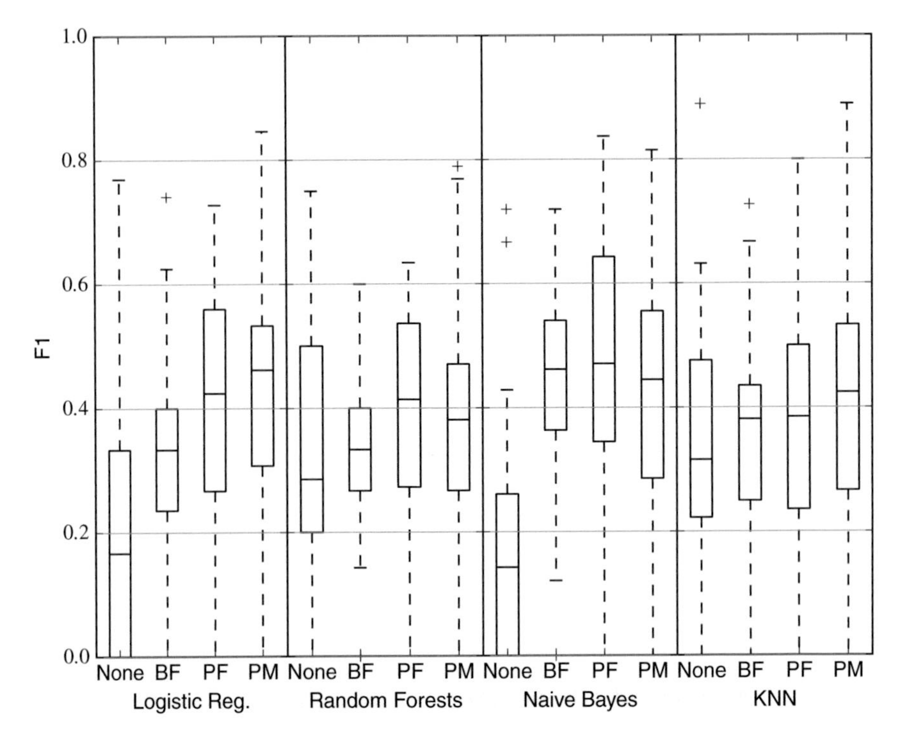

Fig. 3 Predictive performance of cross-project prediction (F-measure)

5 Related Work

Cross-Project Defect Prediction (CPDP) research becomes popular. While initial studies showed the performance was lower than that of within-project defect prediction [10], recent methods achieved the performance improvement. These methods focused on the transformation of cross-project data conforming to a target project data.

Turhan et al. [9] proposed a relevancy filter method based on K-nearest neighbors called Burak-filter. This method selects instances from cross-project data similar to instances in a target project. The performance evaluation used static code features of seven NASA and three Turkish datasets as cross-project data and Naive Bayes Classifier as a prediction method. The authors showed the proposed methods was effective though the use of within-project was still better.

Peters et al. [7] proposed another relevancy filter method. This method focused on the situation where the size of a target project data is small. The proposed method searches relevant cross-project instances based on its own structure. The performance evaluation used static code features of 56 project datasets served on the PROMISE data repository as cross-project data and four prediction methods such as Random Forests. The results suggested the Peters-filter was better than within-project prediction and Burak-filter for small projects.

Ma et al. [5] proposed a method that weights instances of cross-project data while the relevancy filter methods reduced them. Their method utilized Newton's universal gravitation law to calculate the degree of relation between a target project data and cross-project data. They used the same datasets as used in [9] and showed the performance improvement against Burak-filter.

Nam et al. [6] proposed a method based on TCA that learns a transformation which makes cross and target project data more similar. The experiment adopted eight datasets from OSS projects and logistic regression. The results showed significant performance improvement.

The above studies just used single or combined cross-project data provided for experiments. They left a question how to select cross-project data suitable for a target project data among many projects. Herbold [3] tackled this question with two selection methods. The proposed methods relied on the research by He et al. [2] that suggested distributional characteristics of a target project data and suitable cross-project data were correlated. The proposed methods select cross-project data having similar distribution characteristics based on K-nearest neighbors (K-NN) and EM-clustering algorithm respectively. He used 44 versions of 14 software project data having no small projects from the PROMISE repository as cross-project data and five prediction methods such as SVM. The results recommended the method based on K-NN though the performance was still lower than within-project data prediction.

This study focused on relevancy methods proposed in [7] and [9]. The relevancy methods can easily combine with other methods and the size reduction by using them can also reduce the time to perform other CPDP methods. Furthermore, the combination might improve the predictive performance. Therefore, we expected that improving these methods can contribute to progress CPDP research.

6 Conclusion

This study examined the effects of using DBSCAN for a relevancy filter method for cross-project defect prediction. We conducted the experiment with the same setting as [7] to compare the proposed method with Burak-filter and Peters-filter. The result showed that AUC and g-measure supported the use of DBSCAN. Therefore, we concluded the proposed method was better than the others.

This study focused on clustering method for selecting records of the cross-project data for better defect prediction. DBSCAN worked well, but there are many clustering algorithms. Exploring these algorithms is one of future work. The experiment with the other settings must be performed.

Acknowledgments The authors would like to thank the anonymous reviewers for their thoughtful comments and helpful suggestions on the first version of this paper. This work was partially supported by JSPS KAKENHI Grant #25330083 and #15K15975.

References

1. Ester, M., Kriegel, H.P., Sander, J., Xu, X.: A density-based algorithm for discovering clusters in large spatial databases with noise. In: International Conference on Knowledge Discovery and Data Mining, pp. 226–231 (1996)
2. He, Z., Shu, F., Yang, Y., Li, M., Wang, Q.: An investigation on the feasibility of cross-project defect prediction. Autom. Softw. Eng. **19**(2), 167–199 (2012)
3. Herbold, S.: Training data selection for cross-project defect prediction. In: Proceedings of the 9th International Conference on Predictive Models in Software Engineering (PROMISE'13), pp. 6:1–6:10. ACM, New York (2013)
4. Jureczko, M., Madeyski, L.: Towards identifying software project clusters with regard to defect prediction. In: Proceedings of the 6th International Conference on Predictive Models in Software Engineering (PROMISE'10), pp. 9:1–9:10. ACM, New York (2010)
5. Ma, Y., Luo, G., Zeng, X., Chen, A.: Transfer learning for cross-company software defect prediction. Inf. Softw. Technol. **54**(3), 248–256 (2012)
6. Nam, J., Pan, S.J., Kim, S.: Transfer defect learning. In: Proceedings of 35th International Conference on Software Engineering (ICSE 2013), pp. 382–391. IEEE, New York (2013)
7. Peters, F., Menzies, T., Marcus, A.: Better cross company defect prediction. In: Proceedings of 10th IEEE Working Conference on Mining Software Repositories (MSR 2013), pp. 409–418. IEEE, New York (2013)
8. Rahman, F., Posnett, D., Devanbu, P.: Recalling the "imprecision" of cross-project defect prediction. In: Proceedins of the ACM SIGSOFT 20th International Symposium on the Foundations of Software Engineering (FSE'12), pp. 61:1–61:11. ACM, New York (2012)
9. Turhan, B., Menzies, T., Bener, A.B., Di Stefano, J.: On the relative value of cross-company and within-company data for defect prediction. Empir. Softw. Eng. **14**(5), 540–578 (2009)
10. Zimmermann, T., Nagappan, N., Gall, H., Giger, E., Murphy, B.: Cross-project defect prediction: a large scale experiment on data vs. domain vs. process. In: Proceedins of the 7th Joint Meeting of the European Software Engineering Conference and the ACM SIGSOFT Symposium on The Foundations of Software Engineering on European Software Engineering Conference and Foundations of Software Engineering Symposium (ESEC/FSE'09), pp. 91–100. ACM, New York (2009)

A Game Framework Supporting Automatic Functional Testing for Games

Woei-Kae Chen, Chien-Hung Liu, Ping-Hung Chen and Chia-Sheng Hsu

Abstract When developing a computer (mobile) game, testing the game is an important task and takes a large share of the development cost. So far, testing a game's functional features relies mainly on human testers, who personally plays the game, and is thus labor intensive. This paper proposes a method that automates game testing. Since games are usually built on top of game frameworks, the idea is to enhance a game framework with a testing layer, which can execute (playback) test scripts that perform user events and assert the correctness of the game. We design an HTML5 game framework with such a support. In addition, a case study is performed to compare the testing cost of three different methods: writing a test script directly, recording a test script, and testing the game directly by a human tester. The results showed that when repeated testings are necessary, an automatic testing (either writing or recording test scripts) can reduce human cost. Among these three testing methods, recording scripts was the most favored method.

1 Introduction

Software testing plays an important role in software development. Computer (mobile) games, like any other software applications, require thorough testings to ensure their quality, and the testings take a large share of development cost.

W.-K. Chen (✉) · C.-H. Liu · P.-H. Chen · C.-S. Hsu
Department of Computer Science and Information Engineering, National Taipei University of Technology, 1, Sec. 3, Zhongxiao E. Rd., Taipei 10608, Taiwan, Republic of China
e-mail: wkchen@ntut.edu.tw

C.-H. Liu
e-mail: cliu@ntut.edu.tw

P.-H. Chen
e-mail: t100599001@ntut.edu.tw

C.-S. Hsu
e-mail: t101598062@ntut.edu.tw

R. Lee (ed.), *Applied Computing & Information Technology*,
Studies in Computational Intelligence 619,
DOI 10.1007/978-3-319-26396-0_2

For applications with a graphical user interface (GUI), there are testing tools (e.g., [1–5]) that allows a tester to automate the testing of an application's functional features. However, so far, there are no such tools available for games.

Before addressing the proposed game-testing method, we will explain the automatic testing for GUI applications first. A GUI application is usually created by using standard widgets (e.g., Button, TextBox, ComboBox, etc.). Each window (or page) in an application consists of a number of widgets, which interact with the user and also serve as the display, showing important information to the user. These widgets can be made *accessible* (by their vendor) at runtime. That is, when an application is running, a widget on a screen (window) can be searched, queried, and controlled by another test program (or test script). Thus, a test script can perform testing by directing the target application's widgets to perform certain operations and assertions. For example, Microsoft designed Windows controls (widgets) to be accessible and allowed *coded user interface tests* [4]. Similarly, Google provides *uiautomator* [5], which allows testers to create automated test cases for the user interface of Android Apps.

By utilizing the accessibility features of standard widgets, many testing tools (e.g., Robot framework [1], Unified Functional Testing [2], and Selenium [3]) make automated functional testing more efficient. Basically, a tester uses a tool's IDE to create (develop) a *test script*, which is composed of sequences of actions [6, 7]. An action is either an event, which drives the GUI, or an assertion, which affirms the correctness of the application. The test script can be automatically executed by the tool to verify whether the application works correctly. To speedup the development of test scripts, some testing tools (e.g., [2, 3]) also support capture/replay capability, i.e., user interactions to the application can also be captured as test scripts and can be replayed by the tool.

Games are in a sense like GUI applications in that both of them provide interactive visual experience to the users. However, unlike GUI applications, when automating game testing, two major issues arise: (1) accessibility—games do not share a standard set of widgets (visual objects); and (2) timing—for games, timing can be so critical that a slight delay (e.g., 1 ms) may change every thing in the game. A visual object in a game is maybe a man, a monster, a brick, a map, etc., each with completely different shapes, sizes, and user operations (events). Since each game creates its own specialized visual objects, when running a game, there is not a standard way that can access these objects from the test script. Moreover, even if the visual objects can be made accessible, a test script must drive the objects (e.g., sending an event) in precisely the right time. This is difficult to achieve because a slight fluctuation in CPU utilization may alter the execution speed of both the game and test script. When timing is not exactly right, the test script may not always reproduce the same results, which is unacceptable. Therefore, creating a generic game testing tool that can support all different games is highly difficult. Consequently, there are no such tools available for game testing.

Without tools, so far, testing a game's functional features relies mainly on human testers, who personally plays the game and determine whether the game works correctly. This is labor intensive particularly when repetitive testing is necessary. For example, when a developer modifies some source code, a testing that covers all levels (stages) of the game is usually necessary to ensure that there are no regressions.

How to resolve both the accessibility and timing issues? For accessibility, when a game is running, the visual objects of the game must be made accessible so that a test script can drive the game and perform testing. For timing, the test script must be able to perfectly control (synchronize with) the execution of the game so that every test run produces exactly the same results. These issues suggest that, the test script must be run tightly coupled with the game. Note that, instead of creating games from scratch, game developers usually build games on top of game frameworks or engines (e.g., Cocos2D [8] and Unity [9]). Therefore, the idea of this paper is to enhance a game framework with a testing layer, which cooperates with the test scripts to offer accessibility and precise timing control. This is possible because a game framework knows (handles) every visual objects in the game and controls game cycles (a cycle of updating and drawing game objects). Thus, providing such supports is feasible.

We designed an HTML5 game framework that supports automatic functional testing. A game developer can use this HTML5 game framework to develop any games. JavaScript is the programming language of use and the game runs on a browser supporting HTML5. A game developed based on this framework is automatically testable. A tester (sometimes a developer himself) can create a test script either by writing the script directly or by playing the game and capturing the game-play actions. The tester gets the following testing features:

- Test script: a tester can write a test script directly by using a text editor. The test script itself is also written in JavaScript.
- Event: a test script can perform any keyboard or mouse events, and can also request to execute any methods of any objects (e.g., jumping to a special game level or state).
- Assertion: a test script can retrieve any attribute from any objects and assert if the value of the attribute is as expected.
- Capture and replay: in capture mode, user actions can be captured and translated into a test script. A captured test script does not contain assertions. The tester should in general add additional assertions into the test script.
- Timing: timing control is based on game cycle. Each event or assertion is executed in exactly the specified game cycle and is perfectly in-sync with the game. The execution results will always be the same, even if the game runs on a different computer with a different speed.
- Unit testing: QUnit [10], a JavaScript unit testing framework, is incorporated to manage test execution. Each test script is run as a QUnit test case, thereby integrating both unit testing and functional testing in the same framework.

A case study is performed to compare the testing cost of three different methods: writing a test script directly, recording a test script, and testing the game directly by a human tester. The results showed that when repeated testings are necessary, automatic testing (either writing or recording test scripts) can reduce human cost. Among these three testing methods, recording scripts was the most favored method.

The rest of this paper is organized as follows. Section 2 discusses related work. Section 3 presents an HTML5 game framework that supports the proposed testing method. Section 4 reports a case study that compares automated and non-automated testing. A conclusion is given in Sect. 5.

2 Related Work

Software testing, an important part of software development, has been extensively studied in the literature. However, despite computer (mobile) games have gained a huge revenue and playing games becomes an important part of many people's daily life, there are only few researches that studied game testing. None of these researches addressed automatic functional testing. We will discuss each game testing research as follows.

Kasurinen and Smolander [11] analyzed how game developing organizations test their products. They interviewed seven game development teams from different companies and studied how they test their products. The results suggested that game developers focused on soft values such as game content or user experience, instead of more traditional objectives such as reliability or efficiency. Note that user experience testing is lot different from the functional testing studied in this paper. Both testings are important parts of creating a high quality game product. So far, both user experience and functional testing rely mainly on human testers. This paper proposes a way of automating functional testing so as to reduce the overall testing cost.

Zhao et al. [12] studied model-based testing for mobile games. They proposed a general rule model for mobile games by combining the features of the game software and the Wireless Application Protocol (WAP). They also built an adjusted model for generating test cases with the characteristics of mobile game software. A series of patterns were created to facilitate the generation of TTCN-3 test suite and test case. The limitation of this testing method is that it is suitable for games based on WAP only, and it can only test the communication between the game and WAP—functional testing is not supported.

Cho et al. [13, 14] proposed VENUS II system, which supported blackbox testing and scenario-based testing as well as simple load testing for game servers. Massive virtual clients could be created to generate packet loads and test the performance of game servers. Game grammar and game map were used to describe game logics. However, the focus of VENUS II system was the testing of game servers. In contrast, this paper addresses the testing of game clients.

Schaefer et al. [15] presented Crushinator framework, a game-independent framework that could test event-driven, client-server based game applications and automate processes by incorporating multiple testing methods such as load and performance testing, model-based testing, and exploratory testing. The Crushinator allowed a tester to automate large numbers of virtual clients, utilizing different types of test methods, and to find server load limits, performance values, and functional defects. Like, VENUS II system, Crushinator was designed for the testing of game servers, not clients.

Diah et al. [16] studied usability testing for an educational computer game called Jelajah. They discussed an observation method designed for children. Five preschool children aged between five and six years old were selected for their study. By collecting and analyzing data, the levels of effectiveness, efficiency and satisfaction were determined, which in turn gave an overall usability of Jelajah.

To automate functional testing, some game developers create test programs for their own games (e.g., [17]). This is feasible because, given a particular game, both the accessibility and timing issues (see Sect. 2) can be resolved by directly modifying the source code of the game. However, this is very expensive, game dependent, and the support cannot be extended to other games. In contrast, the testing support proposed in this paper is built-in into game frameworks. Therefore, game-independent functional testing can be achieved as long as the game is developed under the proposed game framework.

3 HTML5 Game Framework

This section describes the HTML5 Game Framework, called simply H5GF. H5GF was originally developed in 2013 as a game framework that supports the teaching of object-oriented programming [18]. In 2014, many students (a total of 7 different teams) have successfully used H5GF to create various types of games, indicating that H5GF is suitable for game development. In this paper, we enhance H5GF to support automatic functional testing. H5GF supports three different modes, namely *play* (playing the game), *replay* (running test scripts), and *record* (capturing the player's actions into test scripts) modes.

The architecture of H5GF is shown in Fig. 1. A game developer creates a game object (e.g., a race car) by subclassing `GameObject`. A game object typically uses either `Sprite`, `AnimationSprite`, or `Scene` to display the image of the object (note: an `AnimationSprite` or `Scene` is a composition of one or more

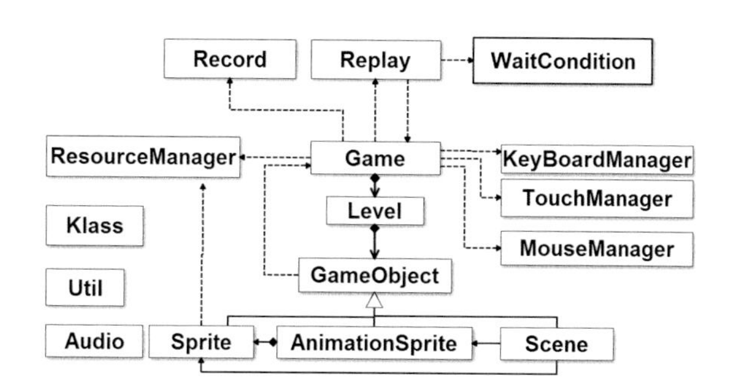

Fig. 1 H5GF class diagram

`Sprite` objects). A `Sprite` object can be rendered on the screen, after specifying its image location (file or URL) and screen position. A game typically has many levels (or stages) and each level uses a lot of different game objects. To create a level, the developer subclasses Level and stores its game objects inside the level object, and then develops the code that processes user events (e.g., keyboard and mouse events) and the code that updates the positions of game objects and draws them.

A game (or a level of a game) developed in H5GF has a life cycle shown in Fig. 2. First, the resources (images and audio clips) needed for showing the "loading progress" are loaded. Then, simultaneously, the game resources are loaded and the loading progress is shown. After that, the game objects in the game are initialized. Then the game enters the main game loop, periodically updating and drawing game objects. Finally, when the game ended, the resources are released (Teardown). A game is developed by providing the code (function) for each activity shown in Fig. 2. In particular, most game logic resides in Update and Draw functions, which control the behaviors of the game.

H5GF controls the transition from one activity to another. The game developer does not need to write any code for the transitions. In addition, as shown in Fig. 3, H5GF also accepts user events and controls *game cycles* (an update/draw cycle is call a game cycle). The game developer defines the game-cycle frequency by specifying a target FPS (frame per second). H5GF automatically calls the update and draw methods periodically, and calls the corresponding event-handling functions when events are triggered by the player.

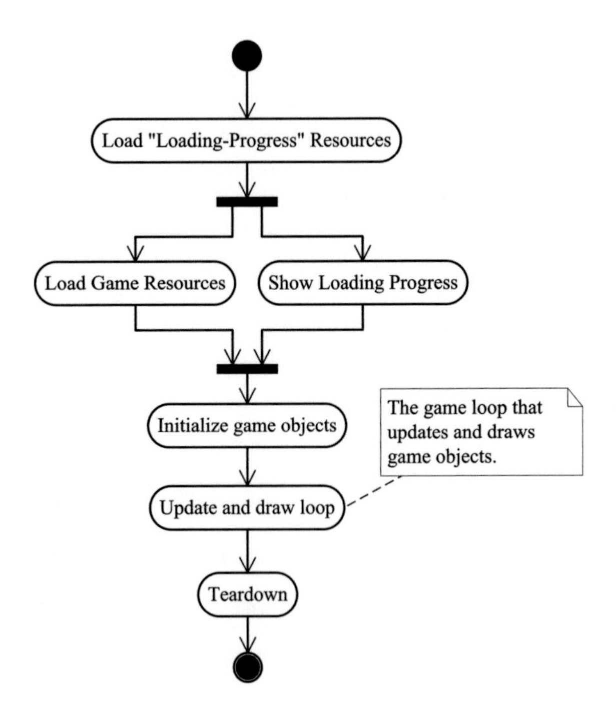

Fig. 2 The life cycle of a game (UML activity diagram)

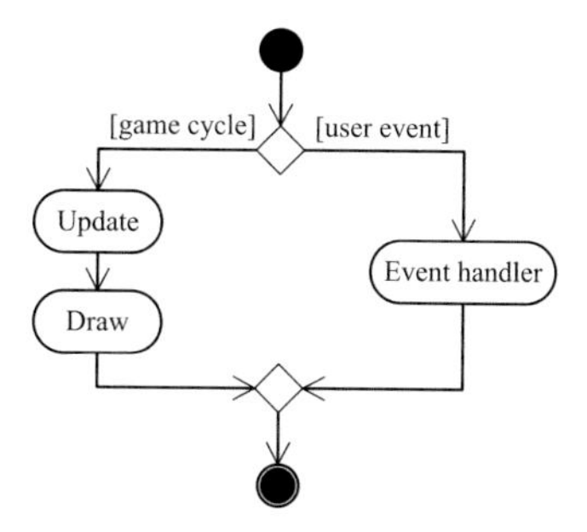

Fig. 3 Game cycle and event processing in play mode

We now describe the testing support of H5GF. The framework uses QUnit [10] to manage test cases and uses three classes, `Record`, `Replay`, and `WaitCondition` to support testing (Fig. 1). In replay (testing) mode, QUnit executes test cases and reports the results. Figure 4 shows a simple test script designed for a well-known game *Bomberman* (the goal of each game level is to

```
QUnit.module("Bomberman Test", {
  setup: function()
  {
    Framework.Replay.start();
  },
  teardown: function()
  {
    Framework.Replay.stop();
  }
});

var Walk = function(direction,times)
{
  for(var i=0;i<times;i++)
  {
    Framework.Replay.keyDown(direction);
    Framework.Replay.waitFor(
      new Framework.WaitCondition(
      "map.player1.isWalking",false));
    Framework.Replay.keyUp(direction);
  }
}

var PlaceBomb = function()
{
  Framework.Replay.keyDown("Space");
}

QUnit.asyncTest("Kill Monster", function( assert ) {
  Framework.Replay.goToLevel("level1");
  Walk('Down' 1);
  PlaceBomb();
  Walk('Up', 2);
  ...
  Framework.Replay.assertEqual(
      "map.monster[0].isdead", true);
  ...
});
```

Fig. 4 A simple test script

place bombs, destroy obstacles, and kill enemies). Lines 1–10 defines the setup and teardown functions, which starts and stops the execution of the game for each test case. Lines 29–38 defines a test case called "Kill monster." When the test case runs, the game jumps to level 1 (line 30), move the bomberman (line 31), place a bomb (line 32), etc. The functions Walk (line 12) and Placebomb (line 24) are reusable actions that can be used to create complicated testing scenarios.

We now explain how the timing issue (see Sect. 1) is resolved. The class `Replay` (see the Replay statements in Fig. 4) performs replay actions and controls timing. To achieve perfect synchronization, the test script and the game are in fact executed in the same thread. In other words, the game is completely frozen, when a test script statement is in execution (e.g., line 16). The only exception is a *waitFor* statement, which allows the game to continue running until a predefined condition is met (e.g., lines 17–19) or run for a specified number of game cycles (e.g., the statement Framework.Replay.waitFor(30) will allow the game to run for exactly 30 game cycles). Figure 5 shows how H5GF interleaves the execution of the test script and the game. In the beginning of each game cycle, H5GF takes an event (or statement) from the test script, and check if the event is a waitFor statement. If yes, H5GF allows the cycle to run until the waitFor condition ends. If not, the event handler is executed. Thus, the test script can precisely control the game without missing even a single cycle. This is crucial to game testing, since missing a cycle can change everything.

How is the accessibility issue resolved? Since both the game and the test script are developed under H5GF. Through H5GF, the test script can access all the game objects created by the game. For example, lines 35–36 in Fig. 4 assert that the first monster is dead by checking whether the attribute "isdead" of the object "map.-monster[0]" is true. Here, a distinctive JavaScript language feature, called *eval*, is

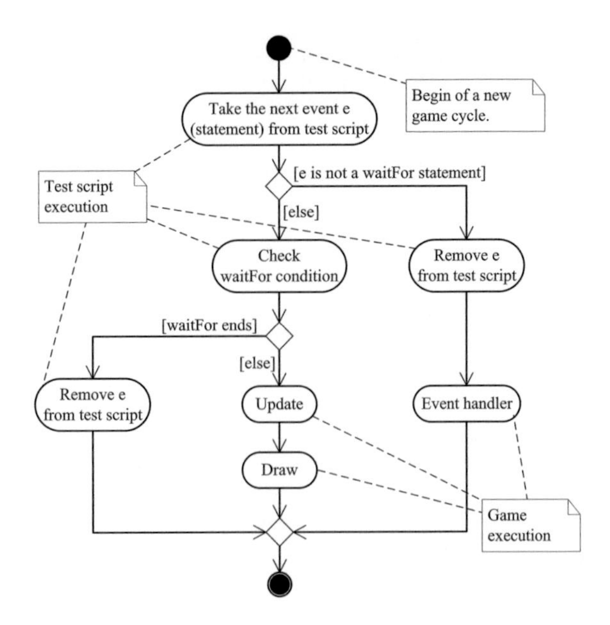

Fig. 5 Game cycle and event processing in replay mode

used to facilitate the access. An eval statement allows JavaScript to treat a string as a JavaScript statement and executes the statement. The assertEqual function (provided by H5GF) uses an eval statement to retrieve the value of "map.monster[0].isdead" and then performs the assertion.

Writing a test script directly can sometimes be cumbersome. Thus, H5GF also provides the ability of capturing (recording) a test script by playing the game. Figure 6 shows how user events are captured in record mode. When an event is triggered, the information of the event is stored and translated into a corresponding test script. The number of game cycles in between two consecutive events are also recorded and translated as a waitFor statement. Figure 7 shows a sample recorded

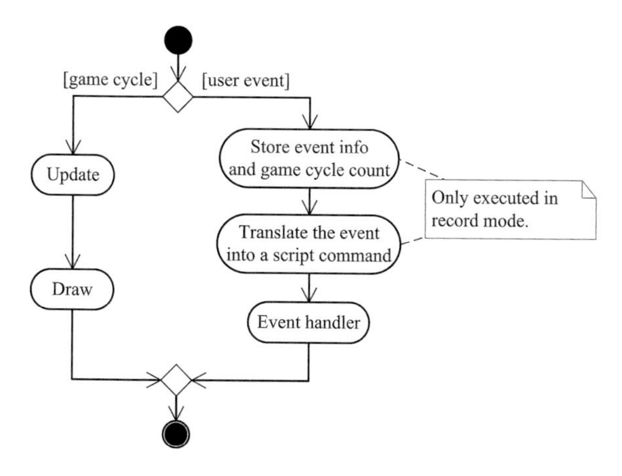

Fig. 6 Game cycle and event processing in record mode

```
1  QUnit.asyncTest("Recorded Script", function(assert) {
2      Framework.Replay.waitFor(124);
3      // Change Level :level1
4      Framework.Replay.waitFor(21);
5      Framework.Replay.mouseClick(457,273);
6      Framework.Replay.waitFor(160);
7      // Change Level :level2
8      Framework.Replay.mouseClick(900,54);
9      Framework.Replay.waitFor(282);
10     // Change Level :level3
11     Framework.Replay.waitFor(33);
12     Framework.Replay.mouseClick(945,48);
13     Framework.Replay.waitFor(375);
14     Framework.Replay.mouseClick(237,524);
15     Framework.Replay.waitFor(343);
16     ...
17     // The assertions are created by the tester
18     Framework.Replay.assertEqual(
19         'player[1].cash',24000,0);
20     Framework.Replay.assertEqual(
21         'player[1].point',0,0);
22 });
```

Fig. 7 A sample recorded test script

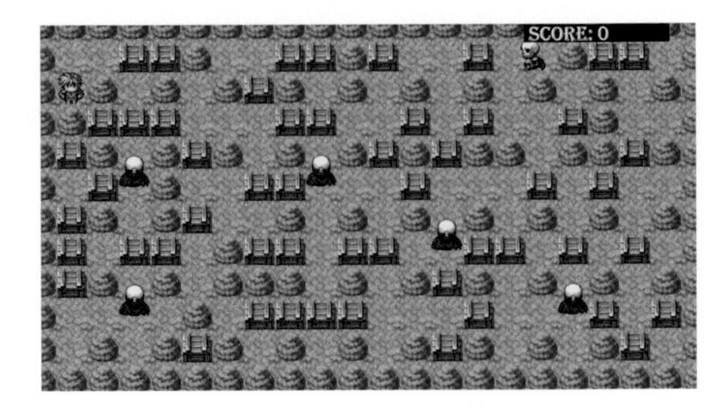

Fig. 8 Bomberman game

script in which mouse click events and waitFor statements are interleaved. Since a recorded script only drives the game, there is no testing at all. Thus, it is important to manually add assertions into the script so that the correctness of the game states can be ensured. Lines 18–21 are such examples. Note that, in comparison to the test script shown in Fig. 4, a recorded script is much more difficult to read. However, capturing a test script is much faster than writing a test script directly. Therefore, it is the tester's decision whether to use capturing function or not.

To verify that H5GF can be used to perform automatic functional testing, we developed a Bomberman game (Fig. 8) and created test scripts for Bomberman by both writing test scripts directly and by recording test scripts. We will report a case study based on Bomberman later in Sect. 4. In addition, we use H5GF to test three different games developed by 2014 students (a flying-shooting game, a monopoly game, and a block-removing game). The test scripts were created by recording player's actions and then enhanced manually with assertions. For all three games, H5GF were able to precisely replay the test script and perform assertions, indicating the support is game-independent.

4 Case Study

The proposed method supports automatic functional testing for games. A test script can be developed either by writing the test script directly or by capturing the test script and then adding assertions manually. Which one is better? In addition, to automate game testing, a tester needs to develop a test script first, which takes time. Therefore, in terms of testing cost (human time), automatic testings do not necessarily outperform human testers. This section reports a case study that compares the cost of the following three different testing methods, namely M1, M2, and M3:

M1 Test a game by writing a test script directly.
M2 Test a game by capturing a test script and then adding assertions manually.
M3 Test a game by a human tester who plays the game personally.

What about maintenance cost? When the contents of a game level is changed (e.g., the arrangement of a game map is changed), an existing test script that was designed for the level may become unusable and needs to be repaired. As described in Sect. 3, there is a trade-off between M1 and M2: using M2 is typically faster than M1, but an M1 script is typically easier to read than an M2 script. So, when taking maintenance (fixing a broken test script) cost into account, which method is better, M1 or M2? The answer to this question may depend on the scale of the change. For example a large-scale change (e.g., the entire game map is changed) may force the entire test script to be rewritten. In these case, it is reasonable to suspect that M2 is better. On the other hand, a small-scale change (e.g., only a small part of the map is changed) may favor M1 over M2, since only a small part of the test script needs modifications. Overall, considering the three testing methods and both their development and maintenance costs, the case study addresses the following research questions:

RQ1 How much cost (human time) do M1, M2, and M3 require?
RQ2 When the contents of a game level is changed, does the scale of the change influences the testing cost.
RQ3 For a large-scale content change in a game level, which method is the best?
RQ4 For a small-scale content change in a game level, which method is the best?
RQ5 Which method is the most favored method?

To answer the above research questions, we designed a case study based on Bomberman (Fig. 8), a game that the player places bombs, destroys obstacles, retrieve special (e.g., power-up) equipments, and kills enemies. Six graduate students were recruited to participate the case study and serve as testers.

Each participant is requested to perform two phases of testing. In the first phase, the participant is given a brand-new game level and requested to test the level thoroughly using all three different methods. The testing includes directing the bomberman to retrieve three different equipments and kill all six enemies to clear the level. For M1, the participant writes the test script step-by-step (moving bomberman, placing bombs, killing enemies, etc.) until the level is cleared. To ensure that the test script contains meaningful assertions, the participant is requested to add assertions that check whether an equipment is indeed retrieved, whether an enemy is indeed killed, and whether the final score is correct. For M2, the participant creates the test script by playing the game and recording the game-play actions, and then adds the same requested assertions into the script. For M3, the participant tests the game simply by playing it. The time that the participant spent on method is recorded.

In the second phase, the maintenance cost is measured. The participant is given a modified game level that is only slightly different from the level given in phase 1 (only one obstacle and one equipment are moved to different places). Again, the participant is requested to test the level thoroughly using all three different methods. Since the test script produced in phase 1 no longer works, for both M1 and M2, the participant is requested to revise (fix) the test script until the level is cleared. For

Table 1 The testing cost of three different methods (*unit* mm:ss)

	M1 time	M2 time	M3 time
Phase 1	51:36	30:52	01:43
Phase 2	16:08	15:07	01:34

M3, the participant re-tests the game simply by playing it. Again, the time that the participant spent on each method is recorded.

The resulting average testing cost (human time) is shown in Table 1. The answer to each of the research questions is listed as follows:

RQ1: How much cost (human time) do M1, M2, and M3 require? Answer: The cost of each method is given in Table 1. For both phases in the case study, M1 was slower than M2, and M2 was slower than M3. Based on the observation of the participants during the case study, the reason that M1 was significantly slower than M2 was because M1 needed to write test-script statements line by line, which was less efficient than recording user actions. The reason that M2 was significantly slower than M3 was because that adding assertions to M2 scripts needed a lot of time. Note that the results did not imply that M3 was the most efficient. If the testing was one-time-only, M3 was indeed the best. However, if the same testing was to be performed repeatedly, both M1 and M2 did not need any additional human time. Therefore, they could be better.

RQ2: When the contents of a game level is changed, does the scale of the change influences the testing cost. Answer: For M3, since a human tester had to re-test everything by hand, the testing costs of a brand-new level (phase 1) was not significantly different from a slightly changed level (phase 2). The story was however different for M1 and M2. Both the testing costs of M1 and M2 reduced significantly from phase 1 to phase 2, indicating that the participants were able to reuse parts of the test scripts developed in phase 1. Therefore, in terms of testing cost, a large-scale change in a game level (e.g., a brand-new level) was more costly than a small-scale change.

RQ3: For a large-scale content change in a game level, which method is the best? Answer: From Table 1, M2 was significantly better than M1 in phase 1. Therefore, we could focus only on the comparison of M2 and M3. In terms of human time, M2 cost was 18 times M3 cost ($30:52 \div 1:43 \approx 18$). Therefore, in case that the same testing was to be repeated for 18 times or more, M2 was the best. On the contrary, if the testing was not repeated for 18 times, M3 was the best.

RQ4: For a small-scale content change in a game level, which method is the best? Answer: From Table 1, M2 was better but only slightly better than M1 in phase 2. In terms of human time, M2 cost was 9.6 times M3 cost ($15:07 \div 1:34 \approx 9.6$). Therefore, if the same testing was to be repeated for 10 or more times, M2 was the best. Note that, for a small-scale change, automatic testings (both M1 and M2) could reuse some test scripts. Thus, for M2 to beat M3, the required number of repetitions was lessened. Moreover, the difference between M1 and M2 was not significant, which suggested that given a different setup (a different game, a different game level, a different difficulty level, etc.), the results could be different.

RQ5: Which method is the most favored method? Answer: After phase 2, the participants were requested to rank the three testing methods according to their own preferences (the most favored method received 1 point, the second 2 points, and the least favored 3 points). Overall, M1, M2, and M3 received 14, 9, and 13 points respectively, indicating that M2 was the most favored testing method.

5 Conclusion

This paper proposes a method of automating functional testing for computer games. An HTML5 game framework was created to address the accessibility and timing issues and offer the testing support. In addition, a case study is performed to compare the testing cost of three different methods: writing a test script directly, recording a test script, and testing the game directly by a human tester. The results showed that automatic functional testing for games is feasible. In addition, the participants of the case study consider recording test scripts (and then adding assertions) as the best testing method. Like testing any other applications, when repeated testing is necessary, automatic game testing can reduce the overall testing cost.

References

1. Robot Framework. http://robotframework.org/. Cited 1 Feb 2015
2. HP. Unified functional testing (UFT). http://www8.hp.com/us/en/software-solutions/unified-functional-automated-testing/. Cited 1 Feb 2015
3. Selenium. Selenium. http://www.seleniumhq.org/. Cited 1 Feb 2015
4. Microsoft. Verifying code by using UI automation. https://msdn.microsoft.com/en-us/library/dd286726.aspx. Cited 1 Feb 2015
5. Google. uiautomator. http://developer.android.com/tools/help/uiautomator/index.html. Cited 1 Feb 2015
6. Memon, A.M., Pollack, M.E., Soffa, M.L.: Hierarchical GUI test case generation using automated planning. IEEE Trans. Softw. Eng. **27**(2), 144–155 (2001)
7. Chen, W.-K., Shen, Z.-W., Tsai, T.-H.: Integration of specification-based and CR-based approaches for GUI testing. J. Inf. Sci. Eng. **24**(5), 1293–1307 (2008)
8. Cocos2D. Cocos2D. http://www.cocos2d-x.org/. Cited 1 Feb 2015
9. Unity. Unity. http://unity3d.com/. Cited 1 Feb 2015
10. QUnit. QUnit. http://qunitjs.com/. Cited 1 Feb 2015
11. Kasurinen, J., Smolander, K.: What do game developers test in their products? In: Proceedings of the 8th ACM/IEEE International Symposium on Empirical Software Engineering and Measurement, Ser. ESEM'14, pp. 1:1–1:10 (2014)
12. Zhao, H., Sun, J., Hu, G.: Study of methodology of testing mobile games based on TTCN-3. In: 10th ACIS International Conference on Software Engineering, Artificial Intelligences, Networking and Parallel/Distributed Computing (SNPD'09), pp. 579–584 (2009)

13. Cho, C.-S., Sohn, K.-M., Park, C.-J., Kang, J.-H.: Online game testing using scenario-based control of massive virtual users. In: The 12th International Conference on Advanced Communication Technology (ICACT 2010), vol. 2, pp. 1676–1680 (2010)
14. Cho, C.-S., Lee, D.-C., Sohn, K.-M., Park, C.-J., Kang, J.-H.: Scenario-based approach for blackbox load testing of online game servers. In: 2010 International Conference on Cyber-Enabled Distributed Computing and Knowledge Discovery, pp. 259–265 (2010)
15. Schaefer, C., Hyunsook, D., Slator, B.M.: Crushinator: a framework towards game-independent testing. In: IEEE/ACM 28th International Conference on Automated Software Engineering (ASE 2013), pp. 726–729 (2013)
16. Diah, N.M., Ismail, M., Ahmad, S., Dahari, M.K.M.: Usability testing for educational computer game using observation method. In: 2010 International Conference on Information Retrieval Knowledge Management, pp. 157–161 (2010)
17. HTML5-benchmark HTML5-benchmark. http://html5-benchmark.com/. Cited 1 Feb 2015
18. Chen, W.-K., Cheng, Y.C.: Teaching object-oriented programming laboratory with computer game programming. IEEE Trans. Educ. **50**(3), 197–203 (2007)

Feature Extraction and Cluster Analysis Using N-gram Statistics for DAIHINMIN Programs

Seiya Okubo, Takaaki Ayabe and Tetsuro Nishino

Abstract In this paper, we elucidate the characteristics of the computer program to play DAIHINMIN, which is a popular Japanese card game with imperfect information. First, we propose a method to extract feature values using the n-gram statistics and a cluster analysis method using the feature values. By representing the program hands as several symbols and representing the order of hands as simplified symbol strings, we obtained the data suitable for feature extraction. Next, we evaluated the effectiveness of the proposed method through computer experiments. In these experiments, we applied our method to ten programs which were used in UECda contest. Finally, we show that our proposed method can successfully cluster DAIHINMIN programs with high probability.

Keywords Imperfect information games · Daihinmin · N-gram statistics · Dendrogram

S. Okubo (✉)
School of Management and Information, University of Shizuoka, 52-1 Yada, Suruga-Ku, Shizuoka 422-8526, Japan
e-mail: s-okubo@u-shizuoka-ken.ac.jp

T. Ayabe
Graduate School of Electro-Communications, University of Electro-Communications, 1-5-1 Chofugaoka, Chofu, Tokyo 182-8585, Japan

T. Nishino
Graduate School of Informatics and Engineering, University of Electro-Communications, 1-5-1 Chofugaoka, Chofu, Tokyo 182-8585, Japan
e-mail: nishino@uec.ac.jp

R. Lee (ed.), *Applied Computing & Information Technology*,
Studies in Computational Intelligence 619,
DOI 10.1007/978-3-319-26396-0_3

1 Introduction

There have been many attempts of letting computer play various games since before. In some games, like Japanese chess or game of go, computer brains are closing in on that of human professionals in terms of strength. These games are perfect information games, in which players are given equal amount of information.

In imperfect information games, on the other hand, information is partially hidden from players. Studies on imperfect information games have also been taking place, where massive data analyses have been conducted using mahjong and other games. One of such imperfect information game is Daihinmin (a.k.a. Daifugo), a Japanese card game equivalent to President. Computer Daihinmin is the practice to let computers play the game, and since 2006, UEC Computer Daihinmin Convention (UECda) has been organized for that purpose [3]. The convention also provides the opportunity to study the algorithm for imperfect information game, where researchers propose various powerful algorithms [1, 4, 5]. Outcomes from these studies are reflected in the participating programs at the Daihinmin convention and steadily improving them year after year.

Among powerful player programs, mainstream programs use Monte Carlo method or other randomized algorithms. In such programs, even the programmers who wrote the programs cannot predict the next hand. Furthermore, due to the high-speed matches characterizing the convention, as well as the absence of professional players in the game, there are no established tactics or styles often observed in popular games.

The purpose of this study is to extract feature quantities of Daihinmin programs. Here, we propose an extraction procedure using n-gram statistics and cluster analytical method. Also, by applying the proposed techniques to the programs taking part in UEC Computer Daihinmin Convention, we evaluate the adequacy of the proposed techniques.

2 Preliminary

2.1 N-gram Statisticsn

N-gram statistics is a linguistic model with a focus on the type and emergence ratio of n items of element strings appearing next to a sequence of words or letters. N-gram statistics is normally used to extract idioms and identify authors in the area of natural linguistic processing, but it is also widely applied to extensive areas of study. By utilizing n-gram statistics in game studies, we can expect to extract procedures with different length and acquire established procedures, such as behavioral choices unconsciously made by players, without the necessity to limit situations. For perfect information games like the game of go, etc., n-gram statistics is used in studies to automatically extract established patterns [2]. For Daihinmin, an imperfect information game, no clustering or other detailed analyses have been made so far.

2.2 *Cluster Analysis*

In this study, we use three types of distance concepts to compute distances. Then, we use the Ward's method for clustering. The computation methods for each distance shall be defined as follows:

Manhattan distance

Manhattan distance can be obtained by measuring the distance between two points along orthogonal coordinate axes. It is defined as Eq. 1.

$$d_1(X, Y) = \sum_{i=1}^{n} |x_i - y_i| \tag{1}$$

Euclidean distance

Euclid distance is a distance applicable in Euclid space. It is defined as Eq. 2.

$$d_2(X, Y) = \sqrt{\sum_{i=1}^{n} |x_i - y_i|^2} \tag{2}$$

Chebyshev distance

Chebyshev distance defines the greatest difference between two vectors in a dimension as the distance. It is deigned as Eq. 3.

$$d_\infty(X, Y) = \lim_{x \to \infty} \sqrt[p]{\sum_{i=1}^{n} |x_i - y_i|^p} = \max_i(|x_i - y_i|) \tag{3}$$

In this study, we use Ward's method, one of the most popular approaches, for hierarchical cluster analysis. Here, the hierarchical cluster analysis means the following method:

1. N item are given.
2. Initialize N clusters. Each cluster has one item.
3. Repeat next steps $N - 1$ times.
   (a) Caluculate the distance between each cluster. We describe the distance between X_1 and X_2 as $d(X_1, X_2)$.
   (b) Merge the nearest 2 clusters.

4. Output the result.

The Ward's method is a technique to minimize the summation of squares of distances from each cluster value to the center of its mass and is defined as Eq. 4.

$$d(C_1, C_2) = E(C_1 \cup C_2) - E(C_1) - E(C_2) \tag{4}$$

where

$$E(C_i) = \sum_{X \in C_i} (d(X, c_i))^2. \tag{5}$$

c_i is the center of mass of the cluster C_i. c_i is defined as Eq. 6.

$$c_i = \sum_{X \in C_i} \frac{X}{|C_i|} \tag{6}$$

3 The Computer DAIHINMIN

Computer Daihinmin is a Daihinmin game played by computers. The University of Electro-Communications has been hosting the conventions for Computer Daihinmin, and we utilize the convention's framework in this study.

While there is an extensive amount of local rules in the game of Daihinmin, UEC Computer Daihinmin Convention adopts following important rules:

Game procedure:

The game is played by five players. It uses a total of 53 cards consisting of 13 (A–K) Hearts, Clubs, Spades and Diamonds, added with a Joker. At the beginning of each game, each player is dealt with 10 or 11 cards. Players take turns in clockwise order to discard (play) cards in hand. The player first getting rid of all cards in hand becomes the winner.

Start of game:

The game starts from the player with the Three of Diamond. Discarding (playing): the player in his/her turn either plays (discards) his/her card(s) or passes the turn. If there is no leading card on the board, the player in turn may play any type of card (single, pair, or kaidan [sequence]). If there is a previous play already on the table, the player in turn may play card(s) that beats the previous play.

To close a round:

When all players pass their turn, then the round comes to a close, and the last player who played a hand may start a new round without a leading hand on the board.

Pass:

Players may pass their turn if they do not have any card to play or prefer to pass. Once a player passes his/her turn, s/he shall not have the turn until the round comes to a close.

Eight Enders (8 rule, 8 GIRI):

A round comes to a close when a player plays a hand containing Eight.

Three of Spades:

When the Joker is played as a single card, a player may close the round by playing Three of Spades.

Revolution (KAKUMEI):

When a player plays a set (pair) of four or more cards with the same number or a kaidan (sequence) with five or more cards, a revolution takes place and the strengths of all cards are reversed until the end of the game.

Lock (Tight, SHIBARI):

When a player follows the same suit with the play currently on the board, the round becomes "tightened" by the suit, and all players must follow the same suit until the round is closed.

Special Titles (Rank, MIBUN):

The player who runs out of cards first is the Daifugo, the second is the Fugo, the third is the Heimin, the fourth is the Heimin, the last one is the Daihinmin.

Cards change (Card trade, Despotism):

The Daifugo trade two cards to the Daihinmin. The Fugo trade one card to the Hinmin. How to choose cards to trade is arbitrary. On the contrary, the Daihinmin trade two strongest cards to the Daihugo. The Hinmin trade one strongest card to the Fugo.

4 Proposed Method

Each program plays cards they are dealt in accordance with their algorithms. The order of cards each Daihinmin program plays shall reveal its features. The proposed procedure consists of the following steps:

1. **Generation of feature quantity vector in each program**: by conducting the following three steps on each program to be analyzed, find out the feature quantity of each program.

   (a) **Log collection**: record each hand played by a program as a log during a match. In this study, we recorded logs of cards played by one program only.
   (b) **Generation of symbol strings from hands played by programs**: from the obtained logs, generate symbol strings that represent cards played by the program for feature extraction.
   (c) **N-gram statistics**: from the symbol strings, calculate the symbol generation probability for 1-gram model and transition probability between symbols for 2-gram model. The obtained generation and transition probabilities are treated as feature quantity vectors.

2. **Cluster analysis**: conduct program clustering through the following two steps:

   (a) **Distance calculation**: calculate the distance between each client program based on the feature quantity vectors.
   (b) **Cluster analysis**: analyze clusters based on the calculated distances to conduct clustering of each client programs.
   (c) **Dendrogramatic illustration**: illustrate the cluster analysis result in a dendrogram.

The procedure to generate symbol strings from programs' hands is as follows: because the initial state is not fixed in Daihinmin, simple replacement of a hand played by a program with a string of letters results in an overwhelming number of states. Therefore, by abstracting the program hands to enable expressions with about a dozen symbols and describing the order of hands with simplified symbol strings, we obtained data suitable for feature quantity extraction.

Table 1 shows the generation method of symbol strings. As single and pair hands occur relatively frequently, we categorized them based on the strength of cards. As sequences are relatively rare, they are not categorized according to their strength. Also, in order to clarify the use of special rules, we prepared specific symbols for "Eight Enders" and "Lock".

Basically, our procedure outputs one symbol in Table 1a every time the program plays a card. As for symbols in Table 1b, they are always used in combination with symbols in Table 1a. Symbol "s" comes in front of the symbols in Table 1a, and Symbol "j" comes after the symbols in Table 1a. For example, the output of paired Two including the Joker, which tightens the round, shall be described as "sej".

Table 1 The notation of the submitted cards

(a) Main characters			
	3–7, 9	10–2	8 GIRI
Single	a	b	c
Pair	d	e	f
Sequence		Pass	
g		p	
(b) Additional characters			
Lock		Joker	
s		j	

5 Outline of Computer Experiment

We evaluated the effectiveness of proposed technique through a computer experiment. In the experiment, we apply the technique to ten programs participating in UECda. Log collection was conducted by letting a Daihinmin program used for feature quantity extraction play against four Default programs for 1000 times.

The ten programs to be studied through feature quantity extraction are categorized in following three groups:

- **Fumiya family: Fumiya, Snowl, and Crow** These Daihinmin programs are characterized by the machine learning they and Monte Carlo method. Incidentally, Fumiya, Snowl, and Crow are champions of 2009, 2010, and 2011 UEC Computer Daihinmin Convention respectively.
- **Sokosoko family: Sokosoko, Teruteru, StrategyBot, and Nakanaka** They are Daihinmin programs loaded with heuristic algorithms to express strategies human players may consider during the game. They are known to originate from Hodohodo.
- **Default family: Default, Customized1, and Customized2** Default is the standard Daihinmin program published on the official website of UEC Computer Daihinmin Convention. Customized1 and Customized2 are variations of client programs with changeable parameters, which are tuned in accordance, to a certain extent, with Default [3].

These Daihinmin programs can be roughly ranked according to their strength as follows: Fumiya family > Sokosoko family > Default family.

We conducted two experiments, in one of which Despotism (card exchange) took place while the other had no Despotism, to evaluate the clustering success rate, etc. Because of implementation methods of respective programs and their origins, clusters had to be formed for each group. Therefore, the "correct answer" in this study was to successfully classify the ten Daihinmin programs into three lineages.

6 Games with Despotism

The computer experiment including the Despotism rule was as follows: we repeated the set of actions consisting of the generation of feature quantity vectors of each program and clustering for ten times, then evaluated the success rate of clustering.

Figure 1 shows a sample result of feature quantity vector obtained through the use of 1-gram statistics, and Figs. 2, 3 and 4 show sample clustering results conducted for respective distances. Table 2 shows the percentage of correct answers in the classification of client programs based on respective distance concepts. In all distance calculation methods, we obtained high clustering success rates.

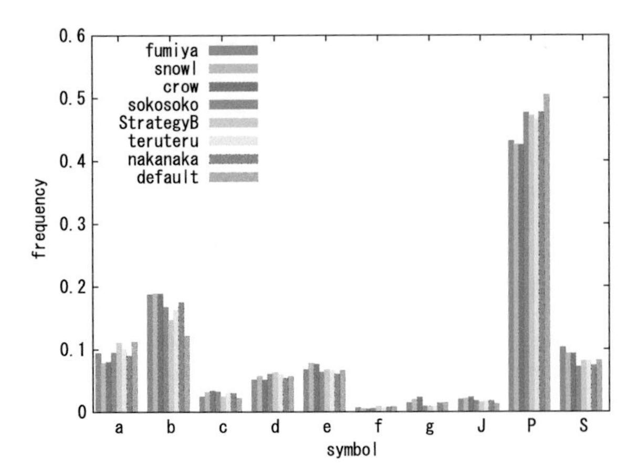

Fig. 1 Feature values (1-gram, with despotism)

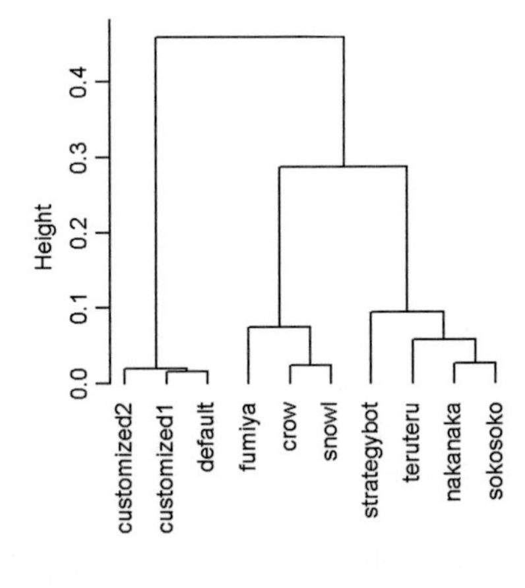

Fig. 2 Dendrogram (Manhattan distance, 1-gram, with despotism)

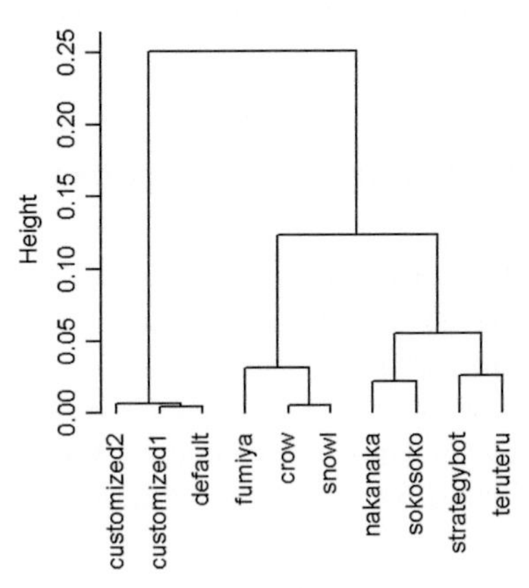

Fig. 3 Dendrogram (Euclidean distance, 1-gram, with despotism)

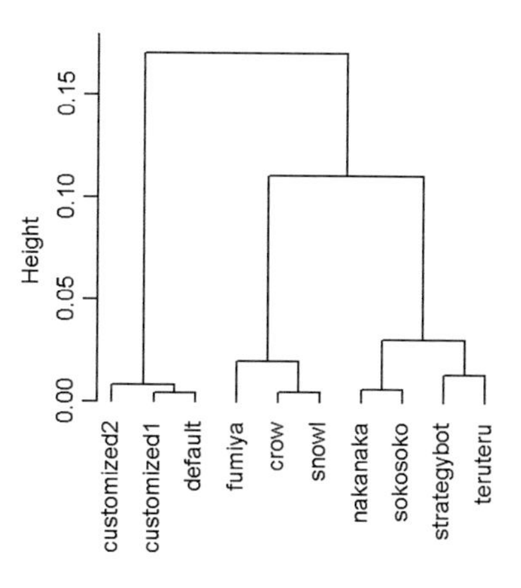

Fig. 4 Dendrogram (Chebyshev distance, 1-gram, with despotism)

Table 2 Success probability(1-gram, with despotism)

	Manhattan distance	Euclidean distance	Chebyshev distance
Success probability	100 %	90 %	100 %

As demonstrated in Fig. 1, feature quantities manifested with significant differences are “s”, “b”, “e” and “p”. In fact, the feature quantities used in the calculation of Chebyshev distance were mostly Symbols “b” (a single card of Ten to Two), “e” (a pair including Ten to Two), and “p” (pass). While these three symbols enabled the classification of the programs into three Daihinmin program families, the occurrence probability of these symbols largely depends on the strength of the Daihinmin program. As strong Daihinmin programs are naturally more likely to assume higher positions, they tend to monopolize powerful cards. Therefore, the occurrence of Symbol “b” (single card, from Ten to Two) and “e” (pairs, from Ten to Two) becomes more frequent, and as they tend to collect advantageous cards, the occurrence of “p” (pass) becomes rare. From these facts, 1-gram analysis of games with Despotism can be said to primarily focus on the strength of the program, and the frequencies of “pass” and “sequence” work as additional features that facilitate the classification.

Figures 5, 6 and 7 show the sample results of clustering of respective distances using 2-gram statistics. Table 3 shows the percentage of correct answers in the classification of client programs based on respective distance concepts. Apart from the low success rate of Manhattan distance, other two methods managed to obtain high clustering success rates.

In the clustering based on Chebyshev distances, the frequency of hands following Symbol “f” (Eight Enders played in pair) was extensively used as the distance that separates the lineages of Daihinmin programs. To be more specific, Default family programs rarely played single Ten to Two (described as Symbol “b”) following Eight Enders in pair (Symbol “f”). Sokosoko family programs often played single Three to Seven and Nine (“a”) following Eight Enders in pair (“f”),

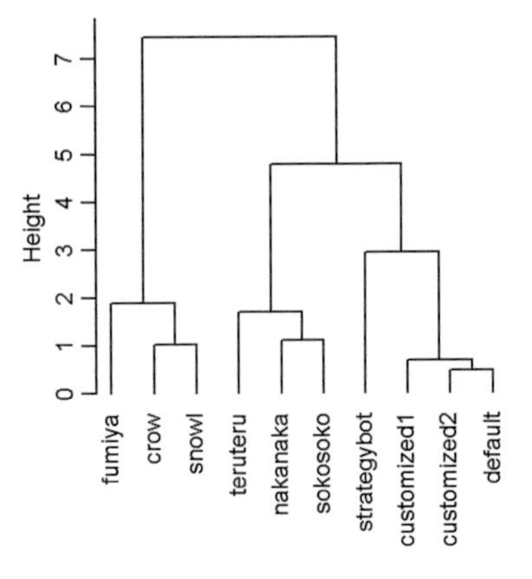

Fig. 5 Dendrogram (Manhattan distance, 2-gram, with despotism)

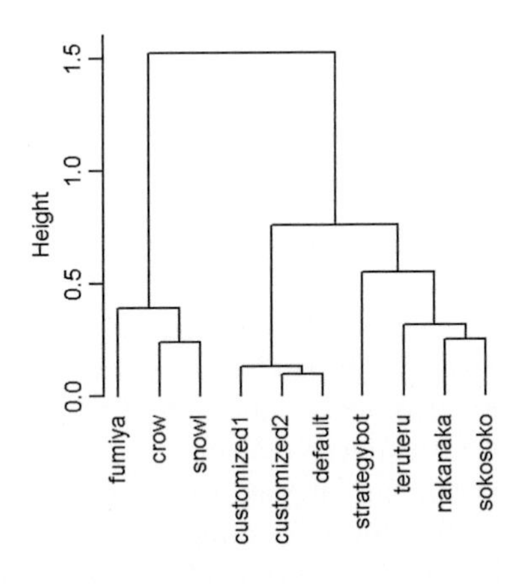

Fig. 6 Dendrogram (Euclidean distance, 2-gram, with despotism)

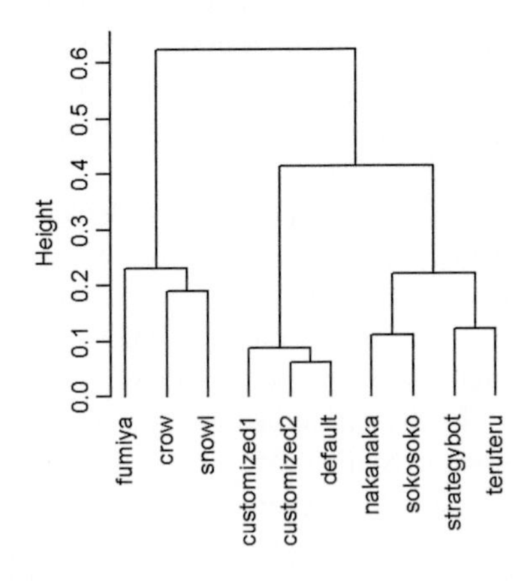

Fig. 7 Dendrogram (Chebyshev distance, 2-gram, with despotism)

Table 3 Success Probability(2-gram, with Despotism)

	Manhattan distance	Euclidean distance	Chebyshev distance
Success probability	60 %	100 %	100 %

while Fumiya family programs frequently play single Ten to Two ("b") following Eight Enders in pair ("f"). Also, the Fumiya family was observed with frequent use of single Ten to Two ("b") following single Eight Ender ("c"), as well as repeated use of Joker in kaidan (sequences). These elements often impacted the computation of Chebyshev distance.

7 Experiment Results Without Despotism

The computer experiment without Despotism rule (card exchange) took place in the following manner: we generated three feature quantity vectors for each eight programs (excluding Customized1 & 2) and clustered the obtained 24 vectors. In this paper, we labeled the three feature quantity vectors as _1, _2, _3. (Namely, 1, 2, 3 are the same program, but they are different trial data to generate feature quantity vectors.) Also, StrategyBot is abbreviated as "stb".

Figure 8 shows a sample result of a feature quantity vector obtained through 1-gram statistics, and Figs. 9, 10 and 11 show the clustering results at respective distances. Figures 12, 13 and 14 show sample clustering results on respective distances obtained through 2-gram statistics.

As the Despotism rule was excluded, there was no bias in the initial state of the dealt cards. The feature quantities of Symbols "a" and "j" in Fig. 8 also support the absence of bias. The result of Default shows the quantity of "b" smaller than that of "e", demonstrating the fact that the program played powerful cards in kaidan

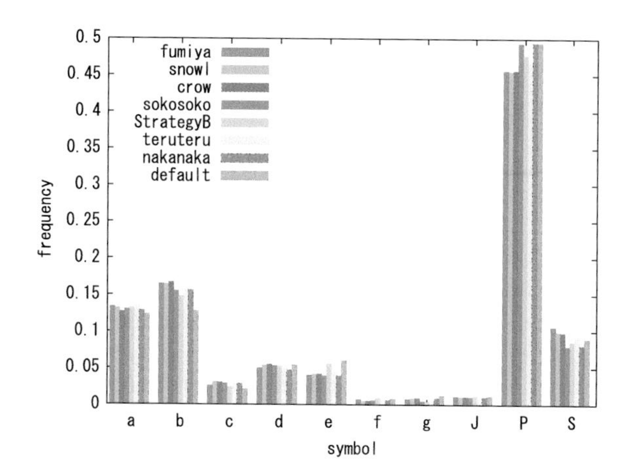

Fig. 8 feature values(1-gram, without despotism)

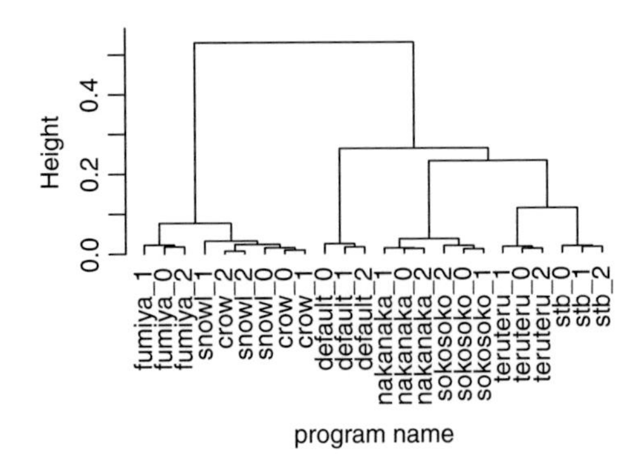

Fig. 9 Dendrogram (Manhattan distance, 1-gram, without despotism)

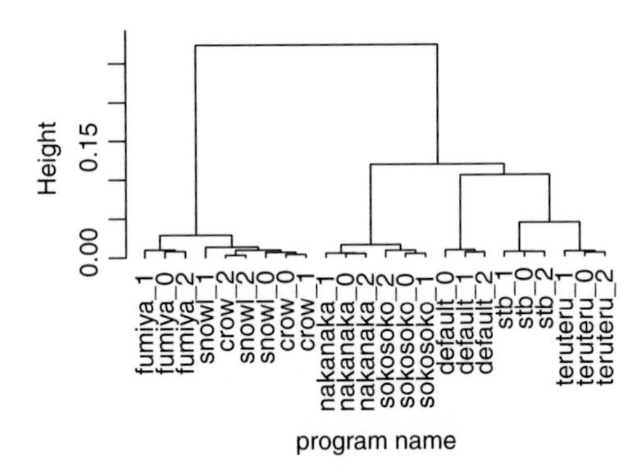

Fig. 10 Dendrogram (Euclidean distance, 1-gram, without despotism)

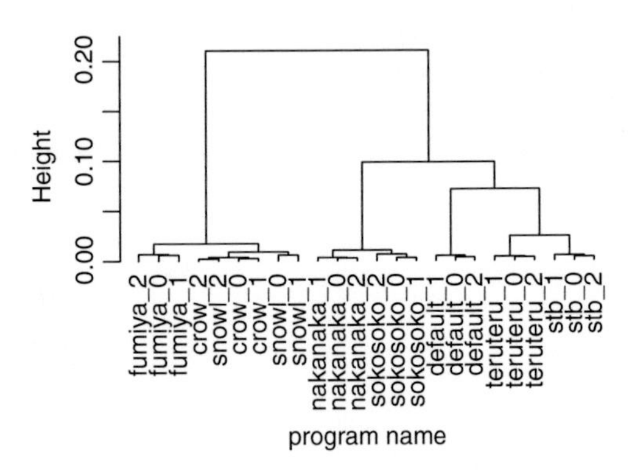

Fig. 11 Dendrogram (Chebyshev distance, 1-gram, without despotism)

(sequence). As the Despotism rule was excluded, the programs are clustered purely based on the hands played. In the clustering results, no difference manifested between Snowl and Crow. This is consistent with the fact that Crow is designed based on Snowl and that their difference only manifests through learning. For other programs, they were respectively clustered and then further clustered in accordance with their lineage. These results support the accuracy of the proposed technique in

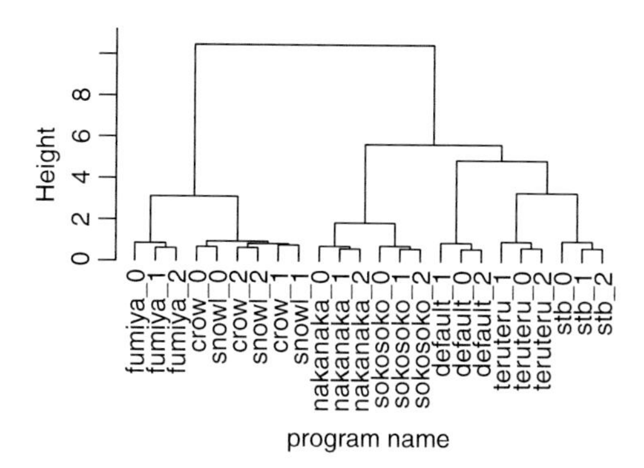

Fig. 12 Dendrogram (Manhattan distance, 2-gram, without despotism)

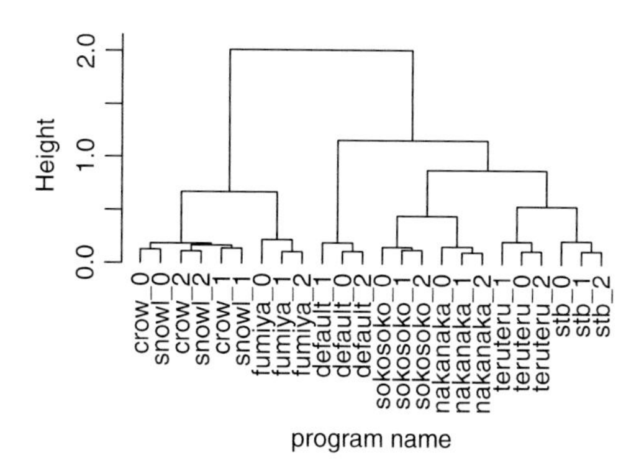

Fig. 13 Dendrogram (Euclidean distance, 2-gram, without despotism)

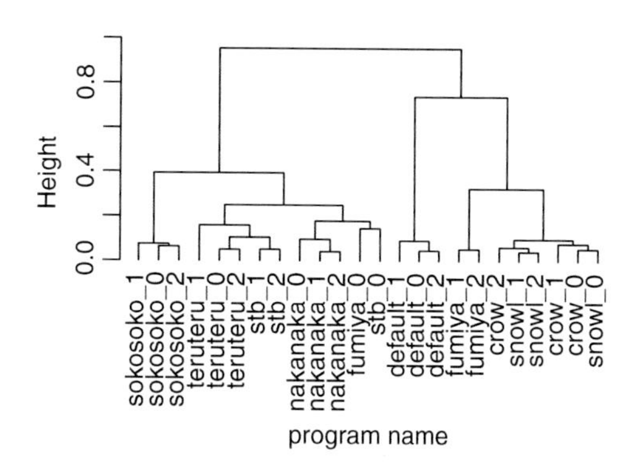

Fig. 14 Dendrogram (Chebyshev distance, 2-gram, without despotism)

clustering programs according to their families (lineages). In Chebyshev distance, however, StrategyBot and Fumiya were falsely perceived. It suggests the limitation of classification based on a single feature quantity without the help of advantages added through the Despotism rule.

8 Discussions

The computer experiment results proved the high accuracy of the proposed technique in clustering features regardless of the presence of Despotism (card exchange rule).

However, in Manhattan distance, the correct answer percentage dropped in 2-gram statistics results. The poor accuracy may be attributable to the method's incompatibility with vectors with higher dimension numbers.

Compared with 1-gram statistics results, results obtained through 2-gram statistics tend to depend on the characteristics of Daihinmin programs rather than on their strength. This may suggest the technique's capability to accurately classify two different Daihinmin programs with equal strength that may emerge in the future.

Clustering results obtained from games with Despotism rule, under Chebyshev distance in particular, suggests the technique's susceptibility to the strength of the programs. While the strength of programs is not the pure characteristics of played hands, however, they still form an important part of the program features. If it is needed to cluster a program together with other programs with equivalent strength and then classify their characteristics, the method may prove to be an effective tool. Although the "correct answer" in this study was to classify programs according to their lineages, we cannot universally conclude that the ability to classify programs according to their family groups is "correct", because the respective clustering methods based on different types of distances respectively evaluated programs based on their own measurements. It is therefore important for a study to choose an adequate log generation method and a distance calculation approach depending on what kind of evaluation it intends to achieve.

In this experiment, characteristics of the play can be identified immediately after 8. Since this play is not affected by other player's play, this play can be used to represent characteristics of Daihinmin programs. We have obtained successful clustering results because the features of the programs can be well observed in this play. On the other hand, in the case of combinations of the other plays, other player's play makes it difficult to obtain good symbol strings to represent features. For further experiments, we have to develop a new method to generate symbol strings to represent the features.

9 Conclusion

In this study, we proposed a technique concerning the feature quantity extraction of Daihinmin program that realizes an objective classification of Daihinmin programs. Through computer experiments, we demonstrated the effectiveness of proposed technique. As the paper's focus was to extract program features and cluster the programs based on the extracted features, we fixed the number of matches at 1000 times and used Default as the fixed opponent program. A future study may include the creation of technique to deal with experiments with unfixed opponents, classification of Daihinmin programs belonging to a new family, feature quantity analysis and clustering of Daihinmin programs with less amount of matches, etc.

References

1. Konuma, S., Honda, T., Hoki, K., Nishino, T.: Application of temporal difference learning to computer Daihinmin. IPSJ SIG Technical Reports. GI, (Game Informatics Research) **2012-GI-27**(1), 1–4 (2012)
2. Nakamura, T.: Acquisition of move sequence patterns from game record database using *n*-gram statistics, pp. 96–105. Game Programming Workshop97 (GPW97) (1997)
3. Nishino, T., Okubo, S.: Computer daihinmin(<special issue> mind games). J. Jpn. Soc. Artif. Intel. **24**(3), 361–366 (2009)
4. Suto, F., Narisawa, K., Shinohara, A.: Development of client "snowl" for computer daihinmin convention. In: Computer DAIHINMIN Symposium 2010 (2010)
5. Suto, F., Shinohara, A.: Development of client for computer daihinmin using the monte carlo method. In: Computer DAIHINMIN Symposium 2009 (2009)

Prediction Interval of Cumulative Number of Software Faults Using Multilayer Perceptron

Begum Momotaz and Tadashi Dohi

Abstract Software reliability is one of the most important attributes in software quality metrics. To predict the number of software faults detected in testing phase, many approaches have been applied during the last four decades. Among them, the neural network approach plays a significant role to estimate and predict the number of software fault counts. In this paper, we focus on a prediction problem with the common multilayer perceptron neural networks, and derive the predictive interval of the cumulative number of software faults in sequential software testing. We apply the well-known back propagation algorithm for feed forward neural network architectures and the delta method to construct the prediction intervals. In numerical experiments with four real software development project data sets, we evaluate the one-stage look ahead prediction interval of the cumulative number of software faults, and compare three data transform methods, which are needed for pre-processing the underlying data, in terms of average relative error, coverage rate and predictive interval width.

Keywords Software reliability · Fault prediction · Multilayer perceptron neural network · Prediction interval · Delta method

1 Introduction

Quantitative software reliability is defined as the probability that software will continuously function during a specified period of time under the well-defined environmental conditions. It is measured from the statistical information of software

B. Momotaz (✉) · T. Dohi
Department of Information Engineering, Graduate School of Engineering Hiroshima University, 1-4-1 Kagamiyama, Higashi-Hiroshima 739-8527, Japan
e-mail: momotaz.2k3@gmail.com

T. Dohi
e-mail: dohi@rel.hiroshima-u.ac.jp

R. Lee (ed.), *Applied Computing & Information Technology*,
Studies in Computational Intelligence 619,
DOI 10.1007/978-3-319-26396-0_4

failures caused by software faults detected in testing phase. In actual software testing phase, it is common to observe the number of software faults or the inter-arrival time between successive fault detection points, which is a vital sign in software reliability. In past, software reliability engineering has provided a number of quantitative methods to measure software reliability and to predict the number of software faults expected in testing [1, 2]. As mentioned above, since the software reliability is defined as a probability, a huge number of stochastic models called software reliability growth model (SRGM) have been developed in the literature. However, no satisfactory SRGM which can be used in every software development project, has been developed from the complex nature of software fault-detection process influenced by human resources spent in the development process.

When we focus on the prediction of the number of software faults in future, it can be regarded as a time-series forecasting problem such as nonlinear regression models. Karunanithi et al. [3–5] are the pioneering works to apply the neural networks to predict the number of software faults in testing phase. Since their seminal contributions, many authors have proposed different neural network architectures and the learning algorithms [6–20]. It is well recognized that artificial neural networks are abstract computing models to describe human brain with many neurons, and can deal with very complex and nonlinear phenomena. The most important point is that artificial neural networks have the capability to learn and represent complex nonlinear relationships. Among them, multilayered feed forward neural networks are simplest, but can possess a universal approximation ability to represent arbitrary nonlinear mapping with any degree of accuracy. In the other engineering field, neural networks are applied to data modeling, time-series prediction, classification, optimization, pattern recognition, and control issues [21, 22].

On the other hand, it should be noted that the neural network-based approach has some drawbacks for application in software reliability assessment. First, the design of neural network architecture, involving the number of neurons in each layer and the number of hidden layers, is arbitrary so that it must be determined carefully through trial-and-error heuristics. Second, the above references with neural network application to predict the number of software faults are based on the deterministic output. In other words, the point prediction of the number of software faults detected at the next testing day is given as an output of complex non-linear functions with trained parameters. Hence, it may be clear that the quantitative software reliability as a probability cannot be evaluated in this framework. The basic but implicit assumption in reliability engineering is that the number of faults is given by a non-negative integer-valued random variable, which is deeply related to the software reliability. Apart from the practical requirement in software engineering, it is often pointed out that the unsatisfactorily low prediction performance arises when neural networks provide ambiguity on the data. In this case, the accuracy of future point forecasts significantly reduces. More specifically, when the training data in neural networks is sparse, the point prediction in neural computing may be less reliable.

In the traditional statistics, the interval estimation is useful to take account of the uncertainty on point estimate itself, though it is difficult to obtain analytically the statistical estimator distribution of the target output. Hwang and Ding [23] and Veaux et al. [24] give the fundamental theory and the learning algorithms to obtain prediction intervals for neural networks, with help of nonlinear regression models. In general, there are three conventional methods to calculate prediction intervals in nonlinear regression problems; delta method, Bayesian method and bootstrap method. Hwang and Ding and Veaux et al. apply the conventional delta method to construct the prediction intervals via neural networks. MacKay [25] considers a Bayesian prediction interval with application to classification problems. Nix and Weigend [26] develop a probabilistic neural network and propose an approximate method to obtain the output probability distribution by means of the mean and variance of the output. On an excellent survey on prediction intervals with neural networks, see Khosravi et al. [27].

In this paper we derive prediction intervals of the cumulative number of software faults detected at each testing day using the multilayer perceptron (MLP) neural network, where the simplest three layers MLP is assumed with the well-known back-propagation algorithm. To predict the number of software faults, we impose a plausible assumption that the underlying fault-detection process obeys the Poisson law with an unknown parameter. Since it is appropriate to input the training data as real number in the conventional MLP neural networks, we propose to apply three data transform methods from the Poisson count data to the Gaussian data: Bartlett transform [28], Anscombe transform [29] and Fisz transform [30]. We apply the delta method [23, 24] to get the prediction intervals of the cumulative number of software faults in the one-stage look ahead prediction. In numerical experiments with actual software development project data, we show that our methods are useful to calculate the prediction intervals on the on-going progress of software debugging process.

2 Neural Network Approach

2.1 Neural Network Architecture

In biological organisms, neural networks are computational metaphor inspired by the brain and nervous system study. They are greatly evaluated by some mathematical models to understand these nervous mechanisms which consist of many simple processing units, called *neurons*. Neurons have interconnections by weights to encode the knowledge of the whole network, which has a leaning algorithm so as to automatically develop internal representations. The most widely used processing-unit models are the *logistic function* and the *sigmoid function*.

The MLP feed forward neural network consists of three types of layers; input layer, hidden layer and output layer. The input layer of neuron can be used to capture the inputs from the external world. The hidden layer of neuron has no communication with the external world, but the output layer of neuron sends the final output to the external world. The main task of hidden layer neurons is to receive the inputs and weights from the previous layer and to transfer the aggregated information to the output layer by any transfer function. This output acts as an input of the output layer. The input layer neuron does not have any computational task. It just receives inputs and associated weights, and passes them to the next layer. On the other hand the feed forward network can work as like the forward direction [4].

The value coming from an input unit (neuron) is labeled by x_i for $i = 1, 2, 3, \ldots, n$ where n is the number of input neurons. There are also two special inputs, bias labeled by x_0 and z_0, which always have the unit values. These are used to evaluate the bias to the hidden nodes and output nodes, respectively. Let H_j ($j = 1, 2, 3, \ldots, m$) be the hidden neuron output, where m is the number of hidden neurons in the hidden layer. Also let w_{ij} be the weight from ith input to jth hidden neuron, where w_{0j} denotes the bias weight form x_0 to jth hidden neuron. The bias input node, x_0, is connected to all the hidden neurons and z_0 is connected to the output neuron. Figure 1 illustrates the MLP neural network architecture under consideration. Each hidden neuron calculates the weighted sum of the input neuron. The hidden node H_j is calculated in Eq. (1).

$$H_j = \sum_{i=1}^{n} w_{ij} x_i + w_{0j} x_0. \tag{1}$$

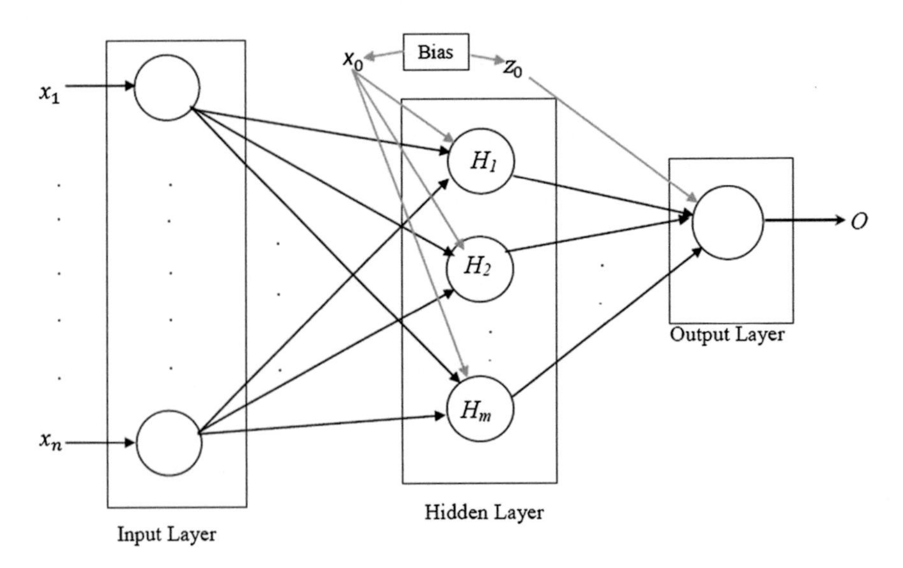

Fig. 1 Architecture of feed forward MLP neural network

After calculating the above equation, we apply the threshold function which is typically either a *step function* or a *sigmoid function*, where the general form of the sigmoid function used in this paper is given by

$$f(H_j) = \frac{1}{1+e^{-H_j}}. \tag{2}$$

In Eq. (2), H_j is the summative and weighted inputs from each hidden neuron. Then, the final output from the output layer is given by

$$O = \frac{1}{1+e^{-\left(\sum_{j=1}^{m} w_{jd} f(H_j) + w_{0d} z_0\right)}}. \tag{3}$$

Here O is the prediction value of the network, w_{jd} is the weight going from the jth hidden neuron to the output neuron, and w_{0d} represents the weight from the bias to the output neuron. In our MLP architecture, all the processing units of input layer are interconnected to all the processing units of the hidden layer, and all the processing units of the hidden layer are interconnected to an output unit, where each weight is associated with each connection.

The general idea on the *back propagation algorithm* [31] is to use the well-known gradient descent method to update the weights so as to minimize the squared error between the network output value and the target output value. The update rules are derived by taking the partial derivative of the error function with respect to the weights to determine each weight's contribution to the error. Then, each weight is adjusted, using the gradient descent, according to its contribution to the error. This procedure is iteratively made for each layer of the network, starting with the last set of weights, and working back towards the input layer, until the desired output is achieved. We propagate the error from an output layer to successive hidden layer by updating the weights. The average sum of squared errors (*SSE*) for the input pattern is represented as

$$SSE^2 = \frac{1}{N-1} \sum_{r=1}^{N} (a_r - O_r)^2, r = 1, 2, \ldots, N, \tag{4}$$

where O_r is the prediction value (point prediction as the neural network output) at rth testing day, a_r is the teaching signals, and N is the prediction period.

2.2 Data Transform

It is common to input real data in the MLP neural network. Since our problem is to predict the number of software faults newly detected at the next testing day, however, the underlying data is integer. On the other hand, it is convenient to treat real number in almost neural network calculations, and to apply the useful property of the Gauss distribution for constructing the prediction intervals approximately (e.g. see [27]). Hence we suppose that the software fault count is described by the Poisson law [1, 2]. In the existing literature, some of authors concern the prediction of the software fault-detection time and handle the real number in their neural network calculations. We apply *Bartlett transform* [28], *Anscombe transform* [29] *and Fisz transform* [30] *as* the most well-known normalizing and variance-stabilizing transforms. The Anscombe's square-root transform (AT) is widely used to pre-process the Poisson data before processing the Gaussian data. Taking the AT, the cumulative number of software fault data can be approximately transformed to the Gaussian data:

$$A_r = 2\sqrt{F_r + 3/8}. \tag{5}$$

In Eq. (5), F_r is the cumulative number of software faults at rth testing day. The AT is a natural extension of the well-known Bartlett transform (BT), and is known as the most fundamental data transform tool in statistics, where BT is defined by

$$B_r = 2\sqrt{F_r + 1/2}. \tag{6}$$

Finally, the Fisz transform (FT) is characterized by the following square root transform as an extension of BT:

$$FT_r = \sqrt{F_r + 1} + \sqrt{F_r}. \tag{7}$$

Table 1 summarizes the data transform methods and their inverse transform methods.

2.3 Delta Method

The delta method used in [23, 24, 27] is a popular method for finding approximations based on Taylor series expansions to the variance of functional of random variables. Basically it is used in the sampling application related to the nonlinear

Table 1 Data transform formulae

Methods	Formulae	
	Data transform	Inverse transform
BT2[18]	$B_r = 2\sqrt{F_r + 1/2}$	$F_r = (B_r{}^2 - 2)/4$
AT[16]	$A_r = 2\sqrt{F_r + 3/8}$	$F_r = (A_r{}^2 - 3/2)/4$
FT[19]	$FT_r = \sqrt{F_r + 1} + \sqrt{F_r}$	$F_r = (FT_r{}^2 + FT_r{}^{-2} - 2)/4$

regression. In the delta method, neural network is interpreted as a nonlinear regression model, which allows us to apply the asymptotic theory based on the Gaussian distribution.

Let ${\delta_r}^T$ is the output gradient vector with respect to gradient values for all output and hidden neurons in the MPL neural network.

$$ {\delta_r}^T = \left[\delta_{O_1}, \delta_{O_2}, \ldots, \delta_{O_r}, \delta_{H_1}, \delta_{H_2}, \delta_{H_3}, \ldots, \delta_{H_j}\right], \quad (8) $$

where δ_{O_r} $(r = 1, 2, 3, \ldots, N)$ are the output gradient of the output layer, δ_{H_j} $(j = 1, 2, 3, \ldots, m)$ are the output gradient of the hidden neuron, m indicates the number of hidden neurons of a hidden layer and a_r is the rth actual value. Then, we have

$$ \delta_{O_r} = O_r(1 - O_r)(a_r - O_r), \quad (9) $$

$$ \delta_{H_j} = \delta_{O_r} w_{jd} f\left(H_j\right)\left(1 - f\left(H_j\right)\right). \quad (10) $$

In practice, the neural network parameters such as pervious weights have to be adjusted by minimizing the average SSE^2. Let Δw_r be the Jacobian matrix in the MLP neural network with respect to all updated weight parameters from an output neuron to hidden neurons. It is computed for all the training samples, where

$$ \Delta w_r = \begin{bmatrix} w_{1d(new)} & w_{11(new)} & w_{21(new)} & \cdots & w_{i1(new)} \\ w_{2d(new)} & w_{12(new)} & w_{22(new)} & \cdots & w_{i2(new)} \\ w_{3d(new)} & w_{13(new)} & w_{23(new)} & \cdots & w_{i3(new)} \\ \cdot & \cdot & \cdot & \cdots & \cdot \\ w_{jd(new)} & w_{1j(new)} & w_{2j(new)} & \cdots & w_{ij(new)} \end{bmatrix}. \quad (11) $$

In Eq. (11), $w_{jd(new)}$ $(d = 1)$ are the new weights of hidden neurons connected to the output neuron, and $w_{ij(new)}$ are the new weights of n input neurons which are connected to m hidden neurons in the hidden layer. These weights, $w_{jd(new)}$ and $w_{ij(new)}$, are given as follows.

$$ w_{ij(new)} = w_{ij} + \alpha w_{ij} + \eta \delta_{H_j} x_i, \quad (12) $$

$$ w_{jd(new)} = w_{jd} + \alpha w_{jd} + \eta \delta_{O_r} O_r. \quad (13) $$

In what follows, we define the prediction intervals (PIs) $[PI_{low}, PI^{up}]$ for the MLP neural network. So the lower limit of PI is given by

$$ PI_{low} = O_r - t_{N-p}^{1-\alpha/2} SSE \sqrt{1 + {\delta_r}^T \left({\Delta w_r}^T \Delta w_r\right)^{-1} \delta_r}. \quad (14) $$

The upper limit is also given by

$$PI^{up} = O_r + t_{N-p}^{1-\alpha/2} SSE \sqrt{1 + \delta_r{}^T \left(\Delta w_r{}^T \Delta w_r\right)^{-1} \delta_r}. \quad (15)$$

In the above equations $t_{\mathrm{N-p}}^{1-\alpha/2}$ is the $(\alpha/2)$—quantile of the student t-distribution function with $(N - p)$ degree of freedom, and p is the number of inputs in the neural network.

The delta method involves a challenging issue to construct PIs, because the computation of the Jacobian matrix (Δw_r) in Eq. (11) and its gradient value $(\delta_r{}^T)$ in Eq. (8) is quite hard with all input data, although the other calculations are relatively simple. Then, in this paper, the Jacobian matrix and the gradient value are calculated and estimated at off-line, although they can be potential sources of computational error for constructing PIs. In addition, the quality of PIs and their optimal values of gradient and Jacobian matrix must be carefully checked to satisfy the convergence condition that the minimum error is achieved at a tolerance level.

3 Numerical Illustration

3.1 Setup

We use four real project data sets cited in [2], DS1, DS2, DS3 and DS4, which are software-fault count data (group data). In these data sets the final testing date and the total number of detected faults are given by (62, 133), (22, 54), (41, 351) and (114, 188), respectively. To find out the desired output via the back propagation algorithm, we need much computation cost to calculate the gradient descent and the Jacobian matrix. Especially, the momentum (α) and the learning rate (η) are the most important turning parameters, where α adjusts the weights and η depends on the convergence speed in the back propagation algorithm. We carefully examine these parameters in pre-experiments with the above data sets and arrive at the following parameter ranges; $\alpha = 0.25$–0.90, $\eta = 0.001$–0.500, where the initial guess of weights ranges from -1.00 to $+1.00$, the number of total iterations in the back propagation run is 1000 and the tolerance level on the minimum error is 0.001.

3.2 Prediction Performance

The prediction performance is evaluated with the software fault count data through the sequential testing, so that we make the one-stage look ahead prediction based on the past observation. More specifically, we use a fixed number of software fault count data as the input data, and predict the next fault count (number of software faults detected at the next testing day) in each shifted prediction point. In the first step, the fault count data detected at each testing day is transformed to the Gaussian data by BT, AT or FT. Next, we input the transformed data into the MLP neural network, where the momentum, learning rate and initial weights are adjusted carefully. By means of the back propagation, we update all the weights and obtain one output which is regarded as a predicted number of software fault counts at the next testing day.

As a prediction performance measure, we introduce the average relative error. Suppose that the observation point is the $(k-1)$-st testing day and that $(k-1)$ software fault counts data are available. For the actual value on the number of software fault counts at the kth testing day, the relative error (RE) is defined by

$$RE_r = \left| \frac{o_r - a_r}{a_r} \right|, \tag{16}$$

where O_r is the prediction value at rth testing time and a_r is the rth actual value (training data). The average relative error takes account of the past history and is defined by

$$AE = \frac{1}{N} \sum_{r=1}^{N} RE_r. \tag{17}$$

Table 2 presents the prediction performance based on AE for four datasets with and without data transform, where the values in round brackets denote the number of input layer neurons and the number of hidden layer neurons, respectively. From this result, the data transform does not lead to the better prediction performance, because AE is independent of the Gaussian distribution property. In Fig. 2, we illustrate the time-dependent behavior of RE and compare four data transform methods (one is no transform) with DS1. From this result, it can be observed that the most stable method is the Fisz transform and the method without transform does not always give the better results in the middle and latter testing phases. In comparison of three transform methods, the most classical Bartlett transform works poorly and cannot give the decreasing trend as the software testing goes on.

3.3 PI Assessment

In order to measure the quality of PIs on the prediction of software fault counts, we need to define two prediction measures called the *PI coverage rate* (PICP) and the *mean prediction interval width* (MPIW) [27]. Assume the significance level as

Table 2 Average relative error in point prediction

Data Set	Average relative error	
	Data transform	AE
DS1 (10, 8)	No Transform	0.3161
	AT	0.6154
	FT	0.4837
	BT	0.7213
DS2 (5, 7)	No Transform	0.6555
	AT	1.1302
	FT	1.1925
	BT	1.7521
DS3 (5, 9)	No Transform	0.4749
	AT	1.9412
	FT	1.2615
	BT	1.4662
DS4 (15, 10)	No Transform	0.7387
	AT	1.7872
	FT	0.6247
	BT	1.2671

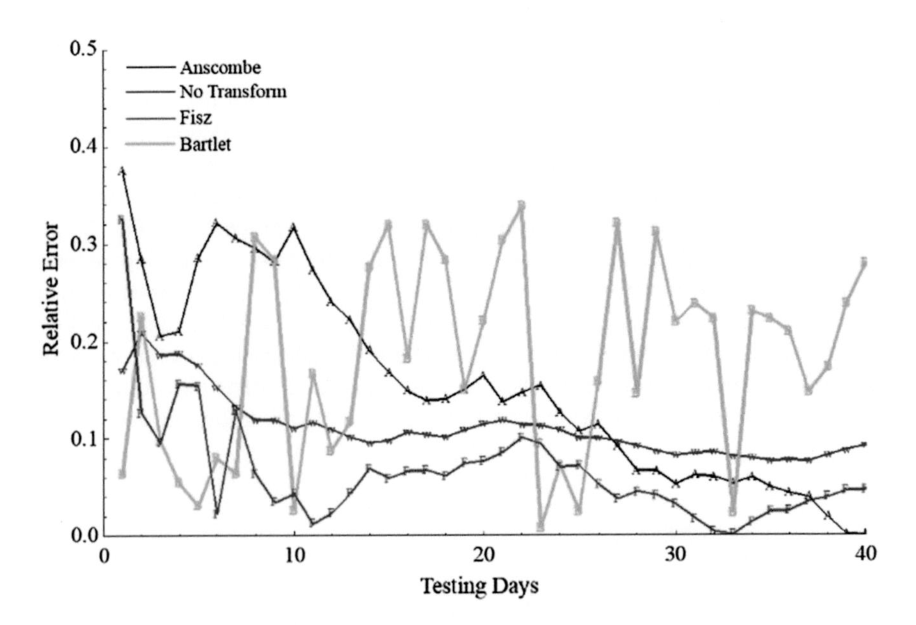

Fig. 2 Time-dependent behavior of relative error (DS1)

95 %. PCIP is the portion of the number of the software fault counts covered by the PI, and is defined by

$$\text{PICP} = \frac{1}{N}\sum_{i=1}^{N} CP_i, \tag{18}$$

where

$$CP_i = \begin{cases} 1, & O_i \in [L_i, U_i], \\ 0, & O_i \notin [L_i, U_i]. \end{cases} \tag{19}$$

L_i and U_i are the lower and upper prediction limits, CP_i is the coverage probability at ith testing day ($i = 1,2,\ldots, k-1$) and O_i is the predicted number of faults. On the other hand, MPIW evaluates the width of PIs, and is defined by

$$\text{MPIW} = \frac{1}{N}\sum_{i=1}^{N} (U_i - L_i). \tag{20}$$

In Table 3 we give the prediction PI measures for four data sets. It can be seen that the data transform does not work to increase PICP in because the corresponding MPIWs are somewhat narrow.

Table 3 Prediction PI measures

Data set	Prediction PI measures		
	Data transform	PICP	MPIW
DS1 (10, 8)	No Transform	96.5204	5056.98
	AT	95.0568	4556.94
	FT	97.9845	3996.81
	BT	95.4485	3031.80
DS2 (5, 7)	No Transform	95.8487	4061.82
	AT	96.4580	3849.68
	FT	96.9587	5391.29
	BT	95.7452	5294.29
DS3 (5, 9)	No Transform	96.8956	4998.96
	AT	97.9805	4597.04
	FT	97.9854	5599.89
	BT	96.9086	4132.62
DS4 (15, 10)	No Transform	96.5898	5469.40
	AT	95.7858	5684.64
	FT	96.5412	6469.40
	BT	96.5188	5226.08

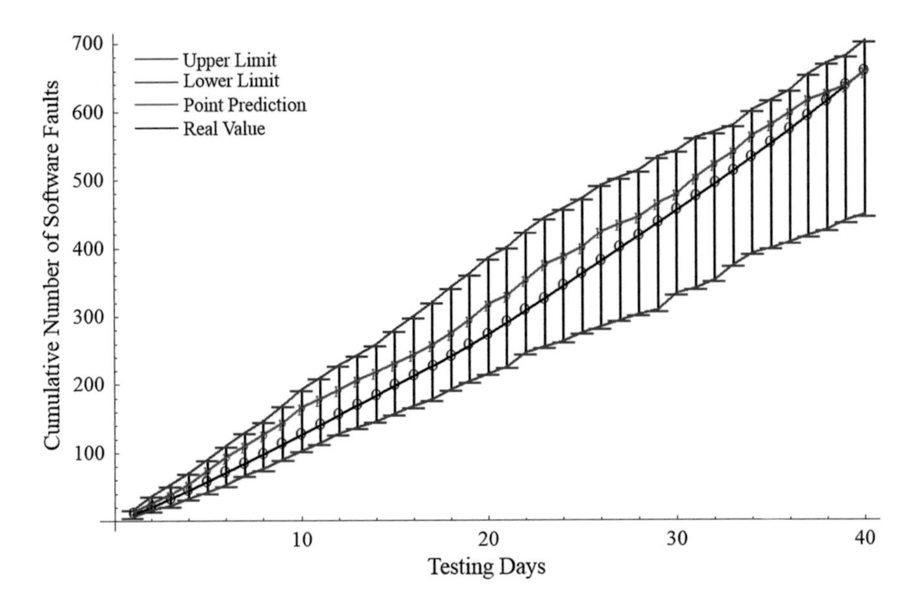

Fig. 3 Anscombe transform (DS1)

For practical usage of PIs, one may understand that the data transform methods are not needed because their associated coverage rates are smaller with wider lengths between the lower and upper prediction limits. However, this intuitive inference seems to be incorrect, because our purpose here is to get the theoretically reasonable PIs which are consistent to the approximate PIs based on the delta methods. This result enables us to know that the PIs suggested in this paper can be used to predict the number of software fault counts under uncertainty, which will experience in the future, and can be useful for the probabilistic inference with subjective significance level controlled by the software test manager.

In Figs. 3, 4, 5 and 6, we depict the sequential prediction results of software fault counts and their PIs with four data transform methods (case with no transform) for the same data sets. It can be found that all the methods cover the one-stage look ahead prediction and the actual data itself within the PIs.

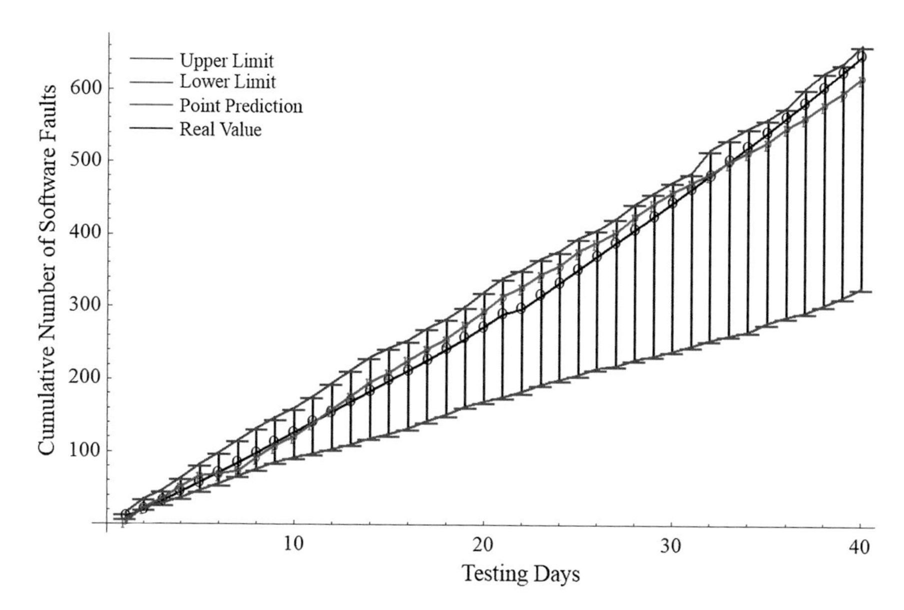

Fig. 4 Fisz transform (DS1)

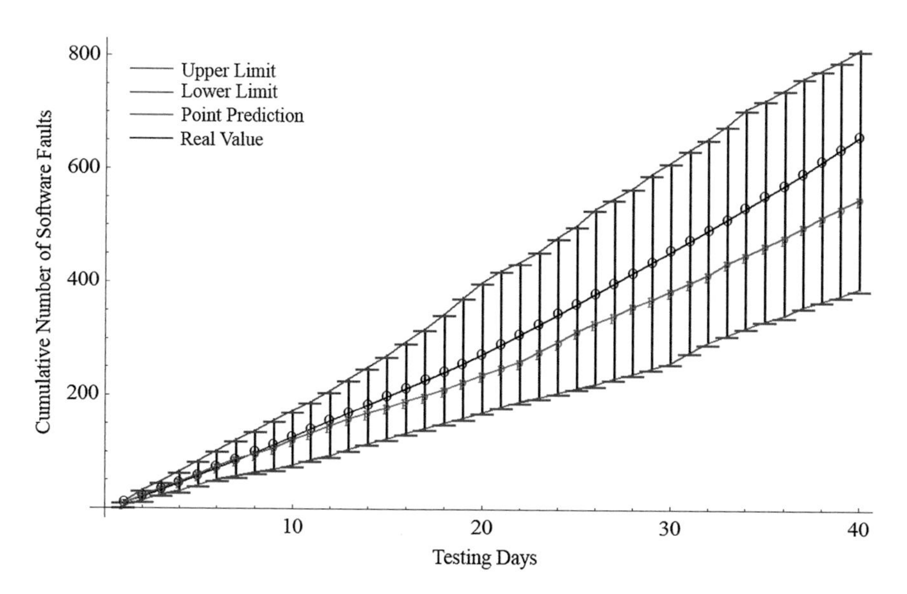

Fig. 5 Bartlett transform (DS1)

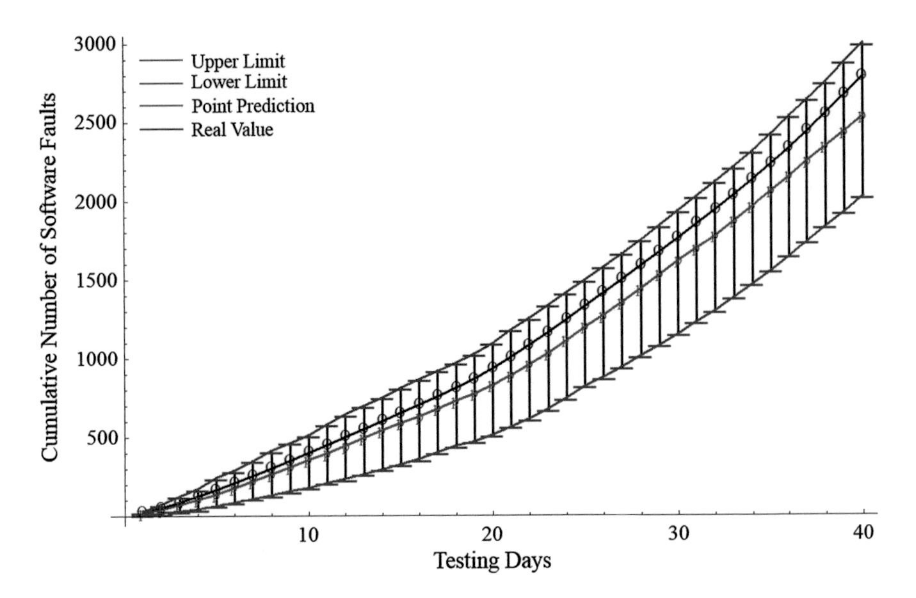

Fig. 6 Case with no transform (DS1)

4 Concluding Remarks

In this paper we proposed prediction intervals of the number of software fault counts using multilayer perceptron neural network, where the data transform methods were applied to calculate them. In numerical experiments with actual software development project data, we evaluated the resulting prediction intervals, and found that those could cover both the point prediction and the actual data in their regions. Although we just gave the real data analysis results based on our proposed methods, these experimental results have to be justified through Monte Carlo simulation in the near future, by comparing with the "real" prediction intervals under the well-defined parametric circumstance.

References

1. Musa, J.D., Iannino, A., Okumoto, K.: Software Reliability Measurement, Prediction, Application. McGraw-Hill, New York (1987)
2. Lyu, M.R. (ed.): Handbook Software Reliability Engineering. McGraw-Hill, New York (1996)
3. Karunanithi, N., Whitley, D., Yashwant, K.: Prediction of software reliability using connectionist models. In: IEEE Transactions on Software Engineering, vol. SE-18, pp. 563–574 (1992)
4. Karunanithi, N., Whitley, D., Yashwant, K.: Using neural networks in reliability prediction. IEEE Softw. **9**, 53–59 (1992)

5. Karunanithi, N., Yashwant, K.: Neural networks for software reliability engineering. In: Lyu, R. (ed.) Handbook of Software Reliability Engineering, pp. 699–728. McGraw-Hill, New York (1996)
6. Khoshgoftaar, T.M., Pandya, A.S., More, H.B.: A neural network approach to predicting software development faults. In: Proceedings of the 3rd International Symposium on Software Reliability Engineering (ISSRE-1992), pp. 83–89. IEEE CPS (1992)
7. Khoshgoftaar, T.M., Lanning, D.L., Pandya, A.S.: A neural network modeling for detection of high-risk program. In: Proceedings of the 4th International Symposium on Software Reliability Engineering (ISSRE-1993), pp. 302–309. IEEE CPS (1993)
8. Khoshgoftaar, T.M., Szabo, R.M.: Predicting software quality during testing using neural network models: a comparative study. Int. J. Reliab. Qual. Saf. Eng. **1**, 303–319 (1994)
9. Khoshgoftaar, T.M.: Using neural networks to predict software faults during testing. IEEE Trans. Reliab. **45**(3), 456–462 (1996)
10. Khoshgoftaar, T.M., Allen, E.B., Hudepohl, J.P., Aud, S.J.: Application of neural networks to software quality modeling of a very large telecommunication system. IEEE Trans. Neural Networks **8**(4), 902–909 (1997)
11. Hochman, R., Khoshgoftaar, T.M., Allen, E.B., Hudepohl, J.P.: Evolutionary neural networks: a robust approach to software reliability problems. In: Proceedings of the 8th International Symposium on Software Reliability Engineering (ISSRE-1997), pp. 13–26. IEEE CPS (1997)
12. Sitte, R.: Comparison of software reliability growth predictions: neural networks vs. parametric recalibration. IEEE Trans. Reliab. **48**(3), 285–291 (1999)
13. Dohi, T., Nishio, Y., Osaki, S.: Optimal software release scheduling based on artificial neural networks. Ann. Softw. Eng. **8**, 167–185 (1999)
14. Dohi, T., Osaki, S., Trivedi, K.S.: Heuristic self-organization algorithms for software reliability assessment and their applications. In: Proceedings of the 11th International Symposium on Software Reliability Engineering (ISSRE-2000), pp. 40–51. IEEE CPS (2000)
15. Cai, K.Y., Cai, L., Wang, W.D., Yu, Z.Y., Zhand, D.: On the neural network approach in software reliability modeling. J. Syst. Softw. **58**(1), 47–62 (2001)
16. Ho, S.L., Xie, M., Goh, T.N.: A study of the connectionist models for software reliability prediction. Comput. Math Appl. **46**(7), 1037–1045 (2003)
17. Thwin, M.M.T., Quah, T.S.: Application of neural networks for software quality prediction using object-oriented metrics. J. Syst. Softw. **76**, 147–156 (2005)
18. Tian, L., Noore, A.: On-line prediction of software reliability using an evolutionary connectionist model. J. Syst. Softw. **77**, 173–180 (2005)
19. Tian, L., Noore, A.: Evolutionary neural network modeling for software cumulative failure time prediction. Reliab. Eng. Syst. Saf. **87**, 45–51 (2005)
20. Su, Y.-S., Huang, C.-Y.: Neural-network-based approaches for software reliability estimation using dynamic weighted combinational models. J. Syst. Softw. **80**, 606–615 (2007)
21. Bose, B.K.: Neural network applications in power electronics and motor drives & mdash: an introduction and perspective. IEEE Trans. Industr. Electron. **54**, 14–33 (2007)
22. Ho, S., Xie, M., Tang, L., Xu, K., Goh, T.: Neural network modeling with confidence bounds: a case study on the solder paste deposition process. IEEE Trans. Electron. Packag. Manuf. **24** (4), 323–332 (2001)
23. Hwang, J.T.G., Ding, A.A.: Prediction intervals for artificial neural networks. J. Am. Stat. Assoc. **92**(438), 748–757 (1997)
24. Veaux, R.D.D., Schumi, J., Schweinsberg, J., Ungar, L.H.: Prediction intervals for neural networks via nonlinear regression. Technometrics **40**(4), 273–282 (1998)
25. MacKay, D.J.C.: The evidence framework applied to classification networks. Neural Comput. **4**(5), 720–736 (1992)
26. Nix, D., Weigend, A.: Estimating the mean and variance of the target probability distribution. In: IEEE International Conference on Neural Networks, vol. 1, pp. 55–60. IEEE CPS (1994)
27. Khosravi, A., Nahavandi, S., Creighton, D., Atiya, A.F.: A comprehensive review of neural network-based prediction intervals and new advances. IEEE Trans. Neural Networks **22**(9), 1341–1356 (2010)

28. Bartlett, M.S.: The square root transformation in the analysis of variance. J. Roy. Stat. Soc. **3** (1), 68–78 (1936)
29. Anscombe, F.J.: The transformation of Poisson, binomial and negative binomial data. Biometrika **35**(3–4), 246–254 (1948)
30. Fisz, M.: The limiting distribution of a function of two independent random variables and its statistical application. Colloquium Mathematicum **3**, 138–146 (1955)
31. Rojas, R.: Neural Networks. Springer, Berlin (1996)

Heuristics for Daihinmin and Their Effectiveness

Yasuhiro Tajima and Kouzou Tagashira

Abstract We show four heuristics for the card game "Daihinmin" and evaluate their effectiveness by some experiments. Our heuristics are implemented on the program called "kou" which is the champion program at UECda-2014 light class. This program uses only heuristics to play Daihinmin. In evaluatio n experiments, we show the strongness of "kou" by matching past champions which contain both of Monte-Carlo algorithm and heuristic algorithm. The result is that our heuristics are effective and our program is as strong as Monte-Carlo algorithm.

1 Introduction

Study on games can be classified into perfect information games and imperfect information games. The perfect information game can be solved by game tree search. The $\alpha - \beta$ method is a famous game tree searching method. In recent years, Monte-Carlo tree search [1] is known as effective method for game tree search. Especially, Go is the most successful application of Monte-Carlo tree search [2].

On the other hand, there is no effective tree searching method for imperfect information games. There is a difficulty that possible states are too many because of imperfectness, then there are too many branches if we create a game tree. But some studies and progress can be found on imperfect information games [3, 4].

"Daihinmin" is a famous card game in Japan. This game is similar to Big-Two which is also a card game mainly played in China [5]. The player is more than 2,

Y. Tajima (✉)
Department of Systems Engineering, Okayama Prefectural University, 111 Kuboki, Soja, Okayama, Japan
e-mail: tajima@cse.oka-pu.ac.jp

K. Tagashira
Graduate School of Computer Science and Systems Engineering, Okayama Prefectural University, 111 Kuboki, Soja, Okayama, Japan
e-mail: cd27029a@cse.oka-pu.ac.jp

R. Lee (ed.), *Applied Computing & Information Technology*,
Studies in Computational Intelligence 619,
DOI 10.1007/978-3-319-26396-0_5

then Daihinmin is a multi-player, imperfect information game. There is a competition of computer Daihinmin called “UECda” [6]. In this annually competition, Monte-Carlo simulation method is an effective move searching method [7] and all champions uses this method in recent years.

In this paper, we show some heuristics for Daihinmin and evaluate whose effects by experiments. The base program to implement our heuristics is called “kou” which is the champion program at light class in UECda-2014 and this does not use Monte-Carlo method. This program is as strong as “snowl” [8] which is the champion program at UECda-2010 and uses Monte-Carlo method. In addition, the time complexity of our program is about one hundred times faster than that of Monte-Carlo method program. This is a new possibility of heuristics on computer Daihinmin.

2 Rules of “Daihinmin”

Daihinmin is a multi-player imperfect information game and uses standard pack of 52 cards with one joker. The number of players is arbitrarily more than 2 but 5 players are the best balance to play. In the following, we assume that the number of players is 5. At first, shuffled almost even number of cards are provided to every player. The 53 cards can not be divided evenly by 5 players, thus three players have 11 cards and two players have 10 cards. The rank of cards has the order $3, 4, 5, \ldots, 9, T, J, Q, K, A, 2$ where “2” is the strongest card and “3” is the weakest. In contrast, the suit “c, d, h, s” is not ordered.

The goal of this game is to get rid of all hands as fast as possible, and the fastest player is called “Daifugo”, the second fastest player is called “Fugo”, the third player is called “Heimin”, the fourth player is called “Hinmin” and the fifth, that is the latest, player is called “Daihinmin.” The round ends when Daihinmin is decided. One round consists of some tricks and a trick is lead by the player who has taken the previous trick. The first trick in the round is lead by the player who has the card “3d.” The player who leads a trick can play arbitrarily and the trick is followed by plays. Every player sits along a playing order. The order is changed after some rounds end.

The “play” of this game can be classified into followings.

- Every single card is a “Single” play.
- Plural cards are one of “Pair”, “Triple” or “Quartet” play if all card ranks of each play are the same. For example, “4s, 4d” is Pair and “Qc, Qh, Qs” is Triple. On the other hand, “7s, 8s” or “6c, 6d, 8h” are not a play.
- The play named “Kaidan” is constructed of more than three cards such that all cards have the same suit and ranks are successive. For example, “4h, 5h, 6h” is Kaidan with three cards, but both of “6c, 7s, 8c” and “8d, Td, Jd, Qd” are not Kaidan.

We call "Single", "Pair", "Triple", "Quartet" or "Kaidan" with the number of cards "types of play." For example, the type of "7s" is Single, the type of "Ts Js Qs Ks" is Kaidan with four cards and the type of "3c, 4c, 5c" is Kaidan with three cards. Every type of play has a specific order of strength as follows.

- On the type of Single, Pair, Triple or Quartet: the rank of the play represents the order. For example, on the type of Triple, "6s, 6c, 6h" is weaker than "7h, 7d, 7c" and it is stronger than "4h, 4d, 4s." On the other hand, "6d, 6c, 6h" has the equal strength. For another example, "2h, 2d" is the strongest on the type of Pair.
- On the type of Kaidan: let α and β be plays whose type is Kaidan. If the highest rank in α is lower than the lowest rank in β then β is stronger than α, or vice versa. If the condition is not satisfied then α and β are incomparable. For example, "3d, 4d, 5d" is weaker than "6c, 7c, 8c" but "4s, 5s, 6s" is incomparable to both of them. Of course, "3d, 4d, 5d" and "6h, 7h, 8h, 9h" are not in the same type then these two plays are incomparable.

The player can select pass or play. Every play must be stronger than the last play in the trick. If the player has no stronger play, he must pass. Once pass is selected, the player can not play until the trick is end. When all players pass, the trick is taken by whom the last play does.

From the shuffled card providing, this game is non-deterministic. This game is zero-sum because one player takes Daifugo then other player can not be Daifugo. Thus, the game "Daihinmin" is multi-player, zero-sum, finite, non-deterministic, imperfect information game.

There are some special rules depend on play.

- Joker can be used as a substitution of any cards. The play "7d, 9d, Joker" is equivalent to "7d, 8d, 9d." "Ac, 2c, Joker" is the strongest Kaidan with three cards. The single Joker is the strongest Single but only "3s" can defeat it.
- "Shibari": When succeeding two plays have corresponding suits, then the trick is in Shibari. If Shibari is initiated, all succeeding plays must also have corresponding suits. In other words, when Shibari begins then possible plays are restricted such that the suits is corresponding until the end of the trick. For example, assume that the last play is "6d" and next player plays "9d", then the successive player can only play "Td", "Jd", "Qd", "Kd", "Ad", "2d" or "joker." Assume that he plays "Ad" then the next player can only play "2d" or "joker."

  For another example, assume that the last play on the trick is "7s, 7d" and the next player plays "Qs, Qd", then the successive player can only play "Ks, Kd", "As, Ad" or "2s, 2d."
- "8": If a play contains a card whose rank is "8", then the trick is terminated. Then a new trick is lead by this player.
- "kakumei" (=revolution): Quartet or Kaidan with more than five cards initiate the "revolution." When it begins, the order of the rank becomes upside down. In the revolution, the rank "3" is the strongest and "2" is the weakest. Twice revolution bring the order into normal.

- Card exchange: When a successive round starts, following card change is done after dealing the cards.
  - Daihinmin (5th at the last round) must submit the best and the second best Single card to Daifugo (1st at the last round), and Daifugo returns arbitrary two cards to Daihinmin.
  - Hinmin (4th at the last round) must submit the best Single card to Fugo (2nd at the last round), and Fugo returns arbitrary one card to Hinmin.

When a round is ended, every player get the following points:

- Daifugo: 5 points
- Fugo: 4 points
- Heimin: 3 points
- Hinmin: 2 points
- Daihinmin: 1 point

After plural rounds, players' ranking is decided by their total points.

3 UECda—The Computer Daihinmin Competition

In the University of Electro-Communications, Japan, computer Daihinmin competition is held every year [6]. The competition has two classes. One is "light class" whose program is restricted to heuristic or some light algorithms. The other is "unlimited class" in which any algorithm is allowed. The competition sets the following environment.

- The number of player is just 5.
- After 100 rounds are played, players' playing order (=sitting position) is changed randomly.
- Total 1000 or 4000 rounds are done and the total of points decides the players' rank.

In UECda-2014, we won the "light class" with the program named "kou" and we introduce our heuristics in this paper.

4 Heuristics

4.1 Algorithm Overview

The following is the overview of the algorithm of "kou" which is the winner of UECda-2014 light class.

1. Finding better combination of plays in the hand such that
   - each of plays has no overlapping and
   - the number of plays is nearly minimum.
2. Select the play which has the highest evaluation value to end the trick. We define the evaluation function for this selection. This function includes some heuristics and special rules to make a priority on plays.
3. If such play is not found, select the play such that "weakness of the hand" is not increase. This means that remaining hand is stronger than the present hand. Usually, the weakest will be played with this criterion.

   The combination of plays in hand is made by the following algorithm.

1. Find all Kaidan whose cards are not any member of Pairs, Triples or Quartets.
2. Let the strongest card and "8" be Single.
3. Find Quartets, Triples and Pairs from the rest of the hand.
4. At last, all the rest cards are Singles.

This algorithm divides player's hand into plays which are not overlapping.

Heuristics used in the evaluation function are as follows.

4.2 Evaluation to End the Trick

The evaluation value is found for each play in the hand. If this value is higher than that of other plays, we think the play is stronger and it tends to end the trick. The evaluation value for a play is calculated by the followings.

1. If the play can end the trick, i.e. there is no possible play which is stronger than the play, then the evaluation value is 100. If the move contains "8" then it is 101. Joker's value is 100 if "3s" is already played, otherwise it is 1.
2. If the play is Kaidan, 100 − (possible Kaidan plays) * (finished player + 1) is the evaluation value. Here, (possible Kaidan plays) is the number of possible Kaidan which can be played by enemies. (finished player) is the number of players who have finished this round.
3. If there are more than 3 players remaining and the play is Pair, Triple or Quartet then $100 - \sum_{i=1}^{m} f_i(n)$ is the evaluation value. Here, m is the number of possible stronger plays which is in enemies' hand. Each i means a play which is in enemies' hand. $f_i(n)$ is found by the following.

   Let n = (the number of stronger cards) − (the number of cards of my play) + (finished player). Then

$$f_i(n) = \begin{cases} 4 & (n \leq 0) \\ 9 & (n = 1) \\ 15 & (n = 2) \\ 24 & (n \geq 3) \end{cases}$$

This means stronger play i takes smaller value of $f_i()$, then the evaluation value will be bigger.
4. Otherwise, i.e. if the move is Single, Pair, Triple or Quartet, 100 − ((stronger total) * 30) is the evaluation value. Here, (stronger total) is the sum of ranks of plays which are in enemies' hands.
5. If these values are less than or equal 0, let it 1.
   If this value of the play is equal to or greater than 95, then we think the play will end the trick. We call this value "strength" of the play. Let m be a play, then the strength of m is denoted by s_m.

4.3 "Shibari" Priority

When the player can initiate Shibari by the play m, the strength of such play s_m is modified to upper limit if the following condition holds.

- After this Shibari, the player has the strongest move.

If this condition does not hold, m will not be played.

4.4 Trick Leading Play

We define play priority for every play m to lead the trick.

The play priority p_m for a play m is calculated by the following.

- If m is Kaidan and contains "8" then $p_m = 9$.
- If m is Joker then $p_m = 3$.
- If $1 \leq s_m \leq 94$ then $p_m = 20 + 2(\textit{rank of m})$. Here, (rank of m) is 1 if play "m" is the weakest and 13 if play "m" is the strongest among plays whose types are the same. For example, a Kaidan "4s 5s 6s" has the (rank of m) of 2 because there exists only weaker play "3s 4s 5s."
- If $95 \leq s_m$ then $p_m = 110 - s_m$.

If p_m is bigger, the play "m" has high priority. If there are some plays whose priority are the same, then the best play is selected from the following order: Kaidan, Triple, Pair and Single.

This value is also used to play when there is no play to end the trick. In such case, combining the following hand weakness, the best play will be selected.

4.5 Strong Play Reservation

We define the hand weakness. The hand weakness is the sum of the following w_m for every play m in the hand.

$$w_m = \begin{cases} 2 & (1 \leq s_m \leq 30) \\ 1 & (31 \leq s_m \leq 60) \\ 0 & (61 \leq s_m \leq 90) \\ -1 & (91 \leq s_m) \end{cases}$$

This value means how the remaining hand is weak. If there is the play m such that s_m is high but the hand weakness is also high, then m will not be played. This is balancing procedure to avoid wasting strong moves.

5 Evaluation Experiments

5.1 Comparing with Past Champions

The following is the list of past light class champions.

- "chibiHana" is the C version of Kishimen_2013 which is the champion of the light class in UECda-2013. This program makes combination of plays whose number is the minimum. The finish search of this program is sophisticated and Shibari strategy is also implemented.
- "Party" is the champion of the light class in UECda-2012. This program tends to have plays which can end the trick. Joker will be used to play weaker cards. Shibari is also considered to end the trick.

Comparing with these two programs, we set the match among the following five programs.

- kou
- chibiHana
- Party
- default
- default

Here, default is the basic strategy program which plays the weakest in the hand. Table 1 is the scores of this setting.

From this result, "kou" is the strongest heuristic algorithm comparing with past two years champions. Especially, "kou" marks high scores at all rounds from 1000 to 10,000.

Table 1 Results among light class champions

Rounds	1000	4000	7000	10000
kou	3660	14954	26178	37413
chibiHana	3493	14136	24817	35460
Party	3271	12698	22208	31536
Default	2324	9198	16017	22900
Default	2252	9014	15780	22691

Table 2 Results against Monte-Carlo programs

Rounds	100	300	500	700	900
Kou	344	920	1481	2021	2528
Beersong	335	999	1541	2213	2853
Paoon	299	891	1570	2209	2893
Crow	246	851	1451	2068	2678
Snowl	276	839	1457	1989	2548

5.2 Comparing with Monte-Carlo Algorithms

Monte-Carlo algorithm is also effective to Daihinmin. In recent years, all champions at "unlimited class" are made with Monte-Carlo simulation. The following is the list of famous Monte-Carlo Daihinmin programs.

- "beersong" is the champion program at UECda-2013.
- "paoon" is the champion program at UECda-2012.
- "crow" is the champion program at UECda-2011 [9].
- "snowl" is the champion program at UECda-2010 [8].

Table 2 is the scores of these four programs and "kou."

From this result, "kou" is as strong as "snowl" which is the early Monte-Carlo algorithm client of Daihinmin. Time complexity of "kou" is very low comparing with any Monte-Carlo programs. We don't have precise data but kou's calculation time is about hundred times faster than other Monte-Carlo programs. Thus, we can say that kou marks the highest scores per second.

5.3 Heuristics Effect

To evaluate the heuristics used in "kou", we have made some modified players to compare with the original kou. The modified player is as follows.

1. Evaluation to end the trick: The original "kou" uses 95 as the threshold value to decide that the play can end the trick or not. So, we prepare w100, w90 and w80 whose thresholds are 100, 90 and 80, respectively.

2. Shibari priority: We prepare programs lock+ and lock−. Here, lock+ plays Shibari if it possible and lock− never plays Shibari.
3. Trick leading play: Two programs are prepared which are weak and single. At the leading of the trick, weak always plays the weakest play in the hand. On the other hand, single always plays the weakest Single.
4. Strong play reservation: The program use_2 does not use "hand weakness" value and "Strong play reservation" heuristic.

Every modified program and the original "kou" are independently matched with the four same enemies. Here, all enemies are the same program which is one of chibiHana, Party or default. For example, we set a test match with one w100 and four chibiHana.

In the followings, all experiments are evaluated by 1000 rounds.

5.3.1 Evaluation to End the Trick

Table 3 shows the score difference between the modified program and the original program.

The program w100 is weaker than the original from this result. On the other hand, w90 and w80 are stronger than the original, thus there will be more optimal threshold value under 95. For chibiHana, the threshold may be between 90 and 95, but it will be under 80 for Party and default. More detailed experiments and analysis are needed to decide the threshold value. Now, we can conclude that 95 is not bad value and this heuristic is effective for various enemies.

5.3.2 Shibari Priority

Table 4 shows the score difference between lock+ or lock− and the original program.

Table 3 Score difference of end the trick heuristic

Modified program enemy	w100 score difference	w90 score difference	w80 score difference
chibiHana	−2132	+403	+210
Party	−767	+663	+970
Default	−825	+15	+121

Table 4 Score difference of Shibari heuristic

Modified program enemy	lock− score difference	lock+ score difference
chibiHana	−1211	−720
Party	−117	−287
Default	−757	−203

Table 5 Score difference of trick leading play heuristic

Modified program enemy	Weak score difference	Single score difference
chibiHana	−165	−1086
Party	−679	−376
Default	−856	−895

From this result, the score of lock− is 1211 less than that of the original. Every difference is minus then we can conclude that both of lock+ and lock− are weaker than the original "kou." Especially, the difference against chibiHana is bigger than the others. This is caused that chibiHana's "end the trick" strategy is more effective than Party and default.

5.3.3 Trick Leading Play

Table 5 shows the score difference about weak and single.

Both modified program weak and single become weaker than the original. Thus we can say that the trick leading heuristic is also effective.

5.3.4 Strong Play Reservation

Table 6 shows the score difference about use_2.

Evaluation of this heuristic's effectiveness needs careful discussion. For chibiHana, this heuristic is effective. But it is not effective for the other two enemies. Obviously, default is the weakest among chibiHana, Party and default. Thus, strong play reserving is effective for the enemies which has complicated algorithm. For a simple enemy, simple play will be effective rather than reserving strong plays.

Table 6 Score difference of strong play reservation heuristic

Modified program enemy	Use_2 score difference
chibiHana	−157
Party	+100
Default	+102

6 Conclusions

In this paper, we introduce the heuristics of "kou" which is the champion program at UECda-2014 light class. The main heuristics are the following four point of view.

- Evaluation to end the trick
- Shibari priority
- Trick leading play
- Strong play reservation

All of these heuristics are effective but Strong play reservation is not effective for some simple algorithm enemies. Moreover, there is possibility that the threshold value or some parameters can be sophisticated by optimization technique or machine learning method. Now, such parameters are provided by hard coding in the program and the values are also selected by heuristic method. When we optimize these values then our program will become more stronger. This is remained for a future study.

In addition, we have shown the strongness of "kou" against past champions and Monte-Carlo algorithms. Especially, "kou" is little stronger than "snowl" which is the champion program at UECda-2010 and it is famous as the turning point that Monte-Carlo algorithm is useful for Daihinmin. It is known that Monte-Carlo algorithm takes much time to play. By our brief counting, heuristic algorithms are one hundred times faster than Monte-Carlo algorithms. Thus, We can show a new possibility of Daihinmin player by heuristic algorithm.

References

1. Browne, C., Powley, E., Whitehouse, D., Lucas, S., Cowling, P.I., Rohlfshagen, P., Tavener,S., Perez, D., Samothrakis, S., Colton, S.: A survey of monte carlo tree search methods, IEEE Trans. Comput. Intell. AI Games **4**(1), 1–49 (2012)
2. Coulom, R.: Efficient selectivity and backup operators in Monte-Carlo tree search. In: Proceedings of the 5th International Conference on Computer and Games, pp. 72–83 (2006)
3. Bowling, M., Burch, N., Johanson, M., Tammelin, O.: Heads-up limit Hold'em Poker is solved. Science **347**(6218), 145–149 (2015)
4. Bonnet, E., Jamain, F., Saffidine, A.: On the complexity of trick-taking card games. In: Proceedings of 23rd IJCAI, pp. 482–488 (2013)
5. Mandai, Y., Hashimoto, T.: A study of big two AI using UCB+, IPSJ Symposium Series **2012** (6), pp. 205–210 (2012) (in Japanese)
6. Daihinmin competition "UECda-2014": http://uecda.nishino-lab.jp/2014/
7. Nishino, J., Nishino, T.: Parallel Monte Carlo search for imperfect information game Daihinmin. In: Proceedings of 5th International Symposium on Parallel Architectures, Algorithms and Programming (PAAP), pp. 3–6 (2012)
8. Sutou, Y., Narusawa, K., Shinohara, A.: Development of Daihinmin client "snowl", 2nd Symposium on UEC Daihinmin Competition (2010) (in Japanese)
9. Konuma, S., Honda, T., Hoki, K., Nishino, T.: Application of temporal difference learning to computer Daihinmin. In: IPSJ SIG Technical Report, 2012-GI-27(1), pp. 1–4 (2012) (in Japanese)

Cluster Analysis for Commonalities Between Words of Different Languages

Jennifer Daniels, Doug Nye and Gongzhu Hu

Abstract The origin of Native American peoples (the Indians) has been a topic of research for many years. Although DNA research has begun to make some progress, there are several competing theories that have yet to be disproved. One of the areas that may reveal the possible relations between Native Americans and other people is through the study of their languages. In this paper, we present a clustering analysis of n = 815 commonly used words in eight different languages including five western languages (English, German, French, Spanish, Italian), an ancient language (Latin), an Asian language (Japanese), and the language of a Native American tribe (Ojibwa). Several similarity measures were established in our clustering analysis using both the word spellings and phonic equivalents (metaphones), in an attempt to discover any underlying relationships between these languages. The results tend to support one of the leading theories describing how Native American tribes, specifically the Ojibwa people, arrived in North America.

Keywords Hierarchical clustering · Similarity metrics · Metaphone coding · Natural language analysis

1 Introduction

The intent of this paper is to demonstrate some techniques that can be used to analyze similarities and differences between languages based on word spelling, letter patterns, and phonic characteristics. The approach we use is, for the most part,

J. Daniels · D. Nye · G. Hu (✉)
Department of Computer Science, Central Michigan University,
Mt. Pleasant, MI 48859, USA
e-mail: hu1g@cmich.edu

J. Daniels
e-mail: danie2je@cmich.edu

D. Nye
e-mail: doug.nye@gmail.com

R. Lee (ed.), *Applied Computing & Information Technology*,
Studies in Computational Intelligence 619,
DOI 10.1007/978-3-319-26396-0_6

to make comparisons between words in different languages with the same meaning. It is our hope that using this approach will help illustrate the potential influence of one language on another as cultures interact, words are shared, and languages evolve.

Most comparative studies on languages derive from the European continent and, due to the all-pervasive effect of the Roman Empire across the area, would be expected to show some similarities. The Roman Empire obviously exerted a profound influence over the region, broadly spreading its culture, institutions, religion, and language. We use eight languages in our study. Four of the languages: Latin, Italian, Spanish, and French, developed in core parts of the Empire while two others, modern English and German, developed in regions adjacent to the core. The remaining two languages in our study, Japanese and Ojibwa, developed far from the influence of the Empire, and would be expected to show much less similarity to the others and perhaps from each other.

In this paper, we describe and evaluate various metrics used in clustering analysis to find similarities between these languages. We hope to discover common elements and characteristics between languages, and to find interesting combinations that may suggest further research opportunities.

2 Related Work

Methods comparing languages based on the percentage of common elements in selected vocabulary has been studied for a long time, pioneered by Swadesh [15]. A variety of techniques have been explored when making these language comparisons. For example, Ethnologue [14] contains statistics on the number of speakers, global location, linguistic affiliations, lexical similarity, etc. for over 7000 languages and dialects. Using Ethnologue's [14] method, which consists of comparing a standardized set of wordlists and counting forms that show similarity in both form and meaning, English was found to have a lexical similarity of 60 % with German and only 27 % with French. Spanish had higher similarities with Italian and French (82 and 75 % respectively). Italian and French were also quite high at 89 %.

Computational methods and algorithms, including clustering techniques [18], have been applied to various problems in natural language processing/analysis [10]. Pereira et al. [11] described a method for clustering English words. In their method, words were represented by the relative frequency distribution, and relative entropy between these distributions was used as the similarity measure for clustering. Lin [9] defined a word similarity measure based on the distributional pattern of words. The similarity measure was then used to construct a thesaurus. Baker and McCallum [1] presented a supervised clustering application for document classification. Words were clustered based on the distribution of class labels associated with each word. A graph clustering algorithm was introduced by Biemann [2] and applied to natural language processing problems.

Heeringa and his colleagues [5] examined different character string metrics for comparison of regional dialects. Johnson and Wichern [6] provided an example of hierarchical clustering for 11 different languages. Serva and Petroni [13] used Levenshtein distance to compare words of the same meaning and averaging on all words in a Swadesh list [16, 17].

3 Methodology

The data set we use for this work contains 8 × 815 words across eight languages. The data was obtained from [4] that contains 815 commonly used words translated into English, French, German, Italian, Japanese, Latin, and Spanish. The Ojibwa translations to these words were added as the eighth language.

Once some appropriate language metrics have been selected, the required computer code was written in order to create several dissimilarity matrices comparing each language. For each matrix, cluster analysis will indicate which languages seem to be the most closely related. These results will also be interpreted to determine which of the competing theories discussing the origins of the Indians seem the most plausible.

3.1 Data Cleaning

As for any statistical data analysis task, the first step of our work was data cleaning. During data cleaning, the following issues were noted:

- The translation database provided multiple words for its English equivalent. Even a relatively simple word such as “cat” had two translations for Spanish (gatto [masculine cat] and gatta [feminine cat]) while French had chat, matou, megere, and rosse. In these cases, the masculine word was chosen (gatto) or the word most closely resembling its English counterpart (chat).
- Some words did not have an easy translation. For example, there is no Latin word for “airplane” since airplanes were not invented until long after the fall of the Roman Empire. In these cases, the entry was simply omitted as missing data.
- Some English words can be ambiguous to other languages. For example, “cabin” and “shack” both have the same translations in Latin (“casa” and “canaba”). It took some time to make certain the correct translations were selected.

Clearly, some of the decisions made during the data cleaning process were subjective but necessary to have a reasonable translation for each word.

3.2 Character String Expressions

For this research, two character string expressions were considered: the actual words as spelled in the original data set, and metaphone encodings.

With regards to metaphone encodings, an interesting question is whether or not words from different languages with the same meaning sound alike or have sounds in common even if they are not spelled the same. It would take someone with linguistic training to accurately make those determinations with a high assurance of accuracy, but the method used here provides a simplistic exploration of the idea.

Metaphone is an algorithm for encoding words by their approximate pronunciation. It was originally developed by Lawrence Philips [12] as an improvement on the Soundex method, which has been used since the 1920s by the U.S. Census Bureau to assist in correcting surname spellings. The metaphone algorithm added support for English words, not just names, and a newer Double Metaphone algorithm added support for more peculiarities of the English language as well as support for a number of other languages. An example of Double Metaphone encodings are shown in Table 1.

The intent here is rather than consider the actual spelling of the words, the phonic sounds can also be compared using the character string metrics discussed in the next section.

3.3 Similarity Measure Metrics of Character Strings

The data set consists of 815 words (of varying lengths) in each of the 8 languages with a total of 815 × 8 = 6520 character strings.

We denote $W = \{W_1, \ldots, W_L\}$ as the set of words in L different languages to be analyzed, where $W_i, 1 \leq i \leq L$ is the set of words in language i. This notation is used in all the algorithms in this paper. $L = 8$ for the data set used in our analysis.

To calculate the similarity (or dissimilarity) of each word pair (w_i, w_j) for $w_i \in W_i$, $w_j \in W_j$, several metrics (Levenshtein distance, Damerau-Levenshtein distance, Euclidean distance) were considered. For this paper, the following statistics, derived from these metrics, were used to create the dissimilarity matrices:

Table 1 Metaphone coding for the word "Moon"

Language	Word	Double metaphone encoding
English	moon	MN
French	lune	LN
Italian	luna	LN
Japanese	tsuki	TSK
German	mond	MNT
Latin	luna	LN
Ojibwa	dbikgiizis	TPKJ
Spanish	luna	LN

(1) Mean Levenshtein distance between original words,
(2) Mean Levenshtein distance between metaphone codes,
(3) Mean Damerau-Levenshtein distance between original words,
(4) Mean Damerau-Levenshtein distance between metaphone codes,
(5) Mean Euclidean distance between each pair of languages, based on frequency of substrings (*n*-grams),
(6) Mean Euclidean distance between each pair of languages, based on length of words.

3.3.1 Levenshtein Distance

The Levenshtein distance [8], also called edit distance, measures the difference between two character sequences. The Levenshtein distance between two words is the minimum aggregate cost of single-character edits (i.e. insertions, deletions or substitutions) required to transform one word into the other. If each of the edit operations has the same cost, the Levenshtein distance would be the minimum number of the edits.

Let A be a finite alphabet set, $I(a)$ be the cost of inserting $a \in A$ into a string, $D(a)$ be the cost of deleting a from a string, and $S(a, b)$ be the cost of substituting a in a string by $b \in A$. For two character strings be $x = x_1x_2...x_n$ and $y = y_1y_2...y_m$, the Levenshtein distance $L(m, n)$ is commonly calculated using the dynamic programming formula.
Initial step:

$$L(0,0) = 0, L(i,0) = \sum_{k=1}^{i} I(x_k), L(0,j) = \sum_{k=1}^{j} D(y_k)$$

Recursive step:

$$L(i,j) = \min \begin{pmatrix} L(i-1,j) + D(y_j) \\ L(i,j-1) + I(x_i) \\ L(i-1,j-1) + S(x_i, y_j) \end{pmatrix}$$

When $I(a) = D(a) = S(a,b) = 1$, the formula become much simpler:

$$\begin{aligned} &L(i,0) = i \\ &L(0,j) = j \\ &L(i,j) = \min(L(i-1,j), L(i,j-1), L(i-1,j-1)) \end{aligned}$$

Each language was compared with its other 7 translated word in this fashion. So, for each specific word, there will be $\binom{8}{2} = 28$ Levenshtein distances resulting from all 8 languages analyzed in pairwise fashion.

The mean Levenshtein distances of language pairs are used for our clustering analysis. We will present the algorithm later in Sect. 4.

3.3.2 Damerau-Levenshtein Distance

Damerau-Levenshtein distance [3] is calculated the same as the Levenshtein distance, except that transposition of two adjacent characters is also allowed as an edit operation. For example, the Levenshtein distance between the words 'abc' and 'acb' would be 2, but the Damerau-Levenshtein distance between "abc" and "acb" would only be 1 since the transposition of "b" and "c" is considered a single operation. Damerau-Levenshtein distance was mostly used for spelling checks as the four edit operations covers over 80 % of human misspellings. It has also been used as a dissimilarity measure for two character strings. Damerau-Levenshtein distance can also be calculated using dynamic programming with a formula similar to the one for Levenshtein distance with one addition recursion for the transposition operations:

$$\begin{aligned}\text{if}\quad i>1 \quad\text{and}\quad j>1 \quad\text{and}\quad x_i = y_{j-1} \quad\text{and}\quad x_{i-1} = y_j\\ L(i,j) = \min(L(i,j), L(i-2,j-2)+1)\end{aligned}$$

3.3.3 Euclidean Distance

The Euclidean distance is a well known distance metric on numeric data:

$$d(x,y) = \sqrt{\sum_{i=1}^{n}(x_i - y_i)^2}$$

To use Euclidean distance as a similarity measure of two character strings x and y, we have to extra numeric values from x and y. One way is to represent the two strings in vector space with the combined characters as the dimensions. Let v be the vector representation of x, $v_a = 1$ if x contains character a, 0 otherwise. Then, the Euclidean distance can be calculated on the two vectors. Another approach is to use the frequency of n-grams [7]. A *n-gram* of string x is a substring of x with length k. For analyzing English words, values of n = 2, 3, 4, 5 are commonly used. In our study, two different word expressions were considered:

- Frequency of n-grams.
- Mean Euclidean distance of lengths of compared words (word by word).

With regards to n-grams, the frequency of all unique n-length substrings (for example 'ee', 'en', 'twi', etc.) were counted. the Euclidean distance was calculated from these n-gram frequencies for each pairwise language comparison.

4 Algorithms

In this section, we shall describe the algorithms for dissimilarity measures using the different metrics mentioned in the previous section. The results of these algorithms are then fed to the hierarchical clustering algorithm to generate dendrograms.

4.1 Levenshtein Distance

The algorithm used to compute the mean Levenshtein distance between two languages is given in Algorithm 1.

4.2 Damerau-Levenshtein Distance

The algorithm for the mean Damerau-Levenshtein distances between the languages is the same as Algorithm 1, except that the calculation of the distance between two words is changed from Levenshtein to Damerau-Levenshtein.

Algorithm 1. Mean Levenshtein distances between languages

Input: $W = \{W_1, \cdots, W_L\}$: word sets in L languages.
Output: $\overline{LevD} = \{\overline{LevD_{ij}}\}$, where $\overline{LevD_{ij}}$ is the mean Levenshtein distance of W_i and W_j.

```
1  begin
2      LevD ← ∅
3      foreach i ≤ L, j ≤ L do
4          LevD_ij ← 0
5      end
6      for 1 ≤ i ≤ L do
7          for 1 ≤ j ≤ L do
8              sum_ij ← 0
9              count_ij ← 0
10             foreach pair (w_i ∈ W_i, w_j ∈ W_j) do
11                 d ← Levenshtein(w_i, w_j)
12                 sum_ij ← sum_ij + d
13                 count_ij ← count_ij + 1
14             end
15             LevD_ij ← sum_ij / count_ij
16             LevD ← LevD ∪ {LevD_ij}
17         end
18     end
19     return LevD
20 end
```

4.3 Distance Using Metaphone Codes

For metaphone codes of the words, the algorithms calculating the mean distances (Levenshtein and Damerau-Levenshtein) between languages are also the same as Algorithm 1 except that the words are converted to their metaphone codes before the distances are computed.

4.4 Euclidean Distance

As mentioned in Sect. 3.3.3, that we use several different representations of the words for calculating the Euclidean distances between words. The algorithm is somewhat different from Algorithm 1 in that it was based on character strings.

The mean Euclidean distances between languages is given in Algorithm 2.

We also calculated the Euclidean distance between languages based on theword lengths, as shown in Algorithm 3.

Algorithm 2. Mean Euclidean distances between languages based on n-grams

Input: $W = \{W_1, \cdots, W_L\}$: word sets in L languages.
Output: $\overline{EucD} = \{\overline{EucD}_{ij}\}$, where $\overline{EucD}_{ij}$ is the mean Euclidean distance of W_i and W_j.

```
1  begin
2      EucD ← ∅
3      foreach i ≤ L, j ≤ L do
4          EucD_ij ← 0
5      end
6      Let G be all n-grams, G ← ∅
7      for 1 ≤ i ≤ L do
8          foreach n ∈ {2,3,4,5} do
9              S ← {n-grams in W_i}
10             count_is ← 0
11             foreach s ∈ S do
12                 count_is ← count_is + 1
13                 G ← G ∪ {s}
14             end
15         end
16     end
17     for s ∈ G do
18         foreach 1 ≤ i ≤ L, 1 ≤ j ≤ L do
19             EucD_ij ← EucD_ij + (Count_is − Count_js)^2
20         end
21     end
22     foreach 1 ≤ i ≤ L, 1 ≤ j ≤ L do
23         EucD_ij ← √(EucD_ij)
24         EucD ← EucD ∪ {EucD_ij}
25     end
26     return EucD
27 end
```

5 Experiments

The algorithms described in Sect. 4 were applied to the data set with the six distance measures, and then the results were fed to the agglomerative hierarchical clustering in R to generate the dendrograms for interpretation of the relationships between the languages.

5.1 Dissimilarity Matrices

For the character expressions (spelling and metaphone) and character metrics previously discussed, six dissimilarity (distance) measures were calculated. An example of the dissimilarity matrix for the first analysis: Mean Levenshtein Distance between languages using the original words is shown in Table 2. Note that the matrix is symmetric.

Algorithm 3. Mean Euclidean distance between languages based on word lengths

Input: $W = \{W_1, \cdots, W_L\}$: word sets in L languages.
Output: $\overline{EucD} = \{\overline{EucD_{ij}}\}$, where $\overline{EucD_{ij}}$ is the mean Euclidean distance of W_i and W_j.

```
1  begin
2      Initialize count_ij, D_ij, EucD_ij, EucD
3      foreach pair of words (w_i ∈ W_i, w_j ∈ W_j) do
4          D_ij ← Dij + (length(w_i) − length(w_j))^2
5          count_ij ← count_ij + 1
6      end
7      foreach i ≤ L, j ≤ L do
8          D_ij = √D_ij
9          EucD_ij = D_ij / count_ij
10         EucD ← EucD ∪ {EucD_ij}
11     end
12     return EucD
13 end
```

5.2 Cluster Analysis

To perform cluster analysis in this context, one must use a clustering algorithm that does not have a predetermined number of clusters (hierarchical) and uses a dissimilarity matrix as the input. In the package **cluster**, used in **R** version 2.15, the Agglomerative Nesting (**agnes**) function performs agglomerative hierarchical clustering using a dissimilarity matrix. There are a variety of options available for

Table 2 Dissimilarity matrix for mean Levenshtein distance

	English	Spanish	Italian	German	Japanese	French	Latin	Ojibwa
English	0							
Spanish	4.726716	0						
Italian	5.056373	4.237745	0					
German	4.805147	5.746324	5.887255	0				
Japanese	5.922604	6.319410	6.491400	6.321867	0			
French	4.330467	4.474201	4.368550	5.441032	6.204182	0		
Latin	5.120541	4.964330	4.890529	5.865929	6.400246	5.001230	0	
Ojibwa	8.130381	8.282903	8.282903	8.279213	8.104551	8.252153	8.304187	0

this function, but the Ward's method was selected as it minimizes the within cluster (internal) variance while maximizing the between cluster variance. The dendogram, which indicates the relationships between the languages, is used to interpret the results.

5.3 Results

The clustering dendrogram using the Levenshtein Distance for the actual word spellings is shown in Fig. 1. It is shown in the figure that when the word's spelling is considered, the Ojibwa language is isolated from the remaining languages. When it does, however, become part of the cluster structure it appears closest to the Japanese language. As expected, the European based languages (English, French, Spanish, Italian) and later Latin and German form a cluster until Japanese joins them later.

When metaphone coding is used, however, the results are much different. The dendrogram resulting from the agens clustering algorithm, using the Levenshtein distance on metaphone coding, is shown in Fig. 2. The dendrogram shows that Ojibwa words are clustered with Japanese. The English and German words cluster together, while Spanish, Italian, French, and later Latin from their own cluster.

Moving on to the Damerau-Levenshtein metric, it can be seen that there is really no difference when compared to the Levenshtein metric. For example, Fig. 3 shows the clustering structure using the Damerau-Levenshtein metric on the original words.

Switching to metaphone encoding, it can be seen that the Damerau-Levenshtein metric (Fig. 4) provides the same results as the Levenshtein metric and is shown in Fig. 2. Again, it can be seen that Japanese is more closely related to Ojibwa than the other European languages.

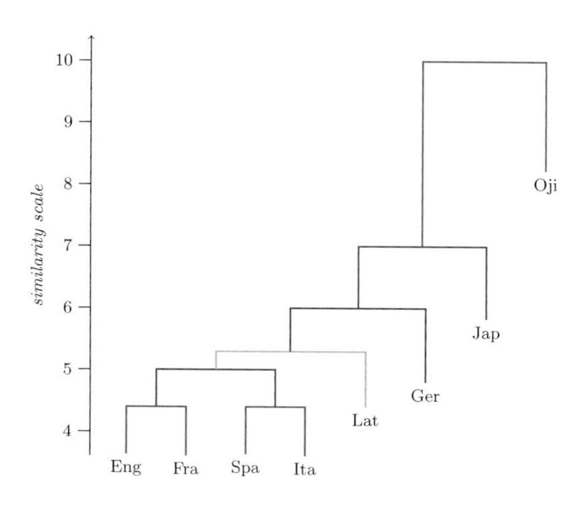

Fig. 1 Dendrogram based on mean Levenshtein distances and actual words (agglomerative coefficient = 0.42)

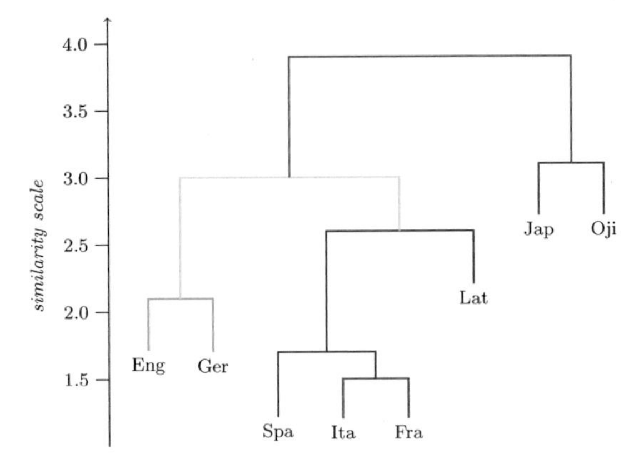

Fig. 2 Dendrogram based on mean Levenshtein distances and metaphones (agglomerative coefficient = 0.42)

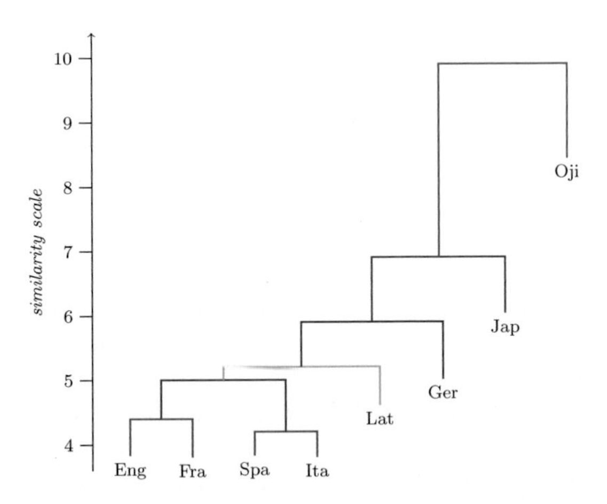

Fig. 3 Dendrogram based on mean Damerau-Levenshtein distances and actual words (agglomerative coefficient = 0.43)

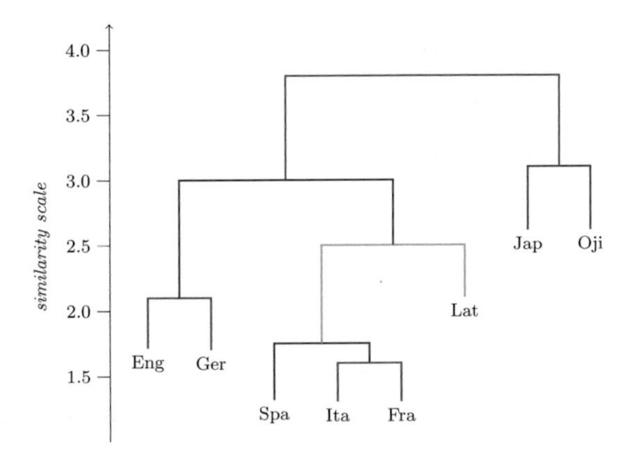

Fig. 4 Dendrogram based on mean Damerau-Levenshtein distances and metaphones (agglomerative coefficient = 0.42)

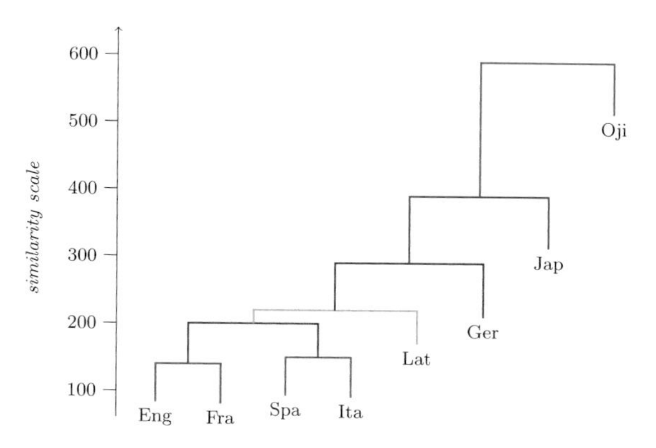

Fig. 5 Dendrogram based on euclidean distances of sub-string frequencies (agglomerative coefficient = 0.56)

Finally, with regards to the Euclidean Distances (dissimilarity matrices 5, 6), all dendrograms appeared the same and provided little insight into the underlying relationships. For example, Fig. 5 is the dendrogram for the Euclidean distance based on n-gram occurrences (dissimilarity matrix 5).

The dendrogram based on Euclidean distance on word length is identical to Fig. 5 and will not be included here.

6 Conclusions

In this paper, we described an application of applying hierarchical clustering algorithm to the data set of words in eight languages. Six dissimilarity metrics were used on the words, examining both their spelling and sound (metaphone codes) to investigate the relationships between these languages. From our investigation, we conclude the following:

(1) Considering word spelling, word length, and frequency of character sub-strings, there is no real similarity between Ojibwa and any other language in this data set.
(2) English/French and Spanish/Italian are spelled similarly with German and Latin joining the cluster later. Japanese and Ojibwa have very different spellings with Ojibwa being the most dissimilar.
(3) When one considers the sound of the language (metaphone), however, there is a relationship between Ojibwa and Japanese. This conclusion tends to lend support to the theory that the Ojibwa came from a Northern Asia origin since their languages appear to sound more alike.
(4) English and German sound alike, while Spanish, Italian, French, and later Latin sound more alike.
(5) The early explorers to North America (Spanish, French, and English) have had little, if any, influence on the evolution of the Ojibwa language. There is very little similarity between these European based languages and Ojibwa.

For future work, more languages (particularly Asian and Pacific languages) may be added in consideration of alternative migration paths for the Indians.

The International Phonetic Alphabet (IPA) is a standardized, character-based representation of the sounds of spoken language, and is intended to include every sound in all human languages. In IPA, there is one character for each distinctive sound. We are also considering to extend this research by considering IPA as a character string expression.

References

1. Baker, L.D., McCallum, A.K.: Distributional clustering of words for text classification. In: Proceedings of the 21st Annual International ACM SIGIR Conference on Research and Development in Information Retrieval, pp. 96–103. ACM (1998)
2. Biemann, C.: Chinese whispers: an efficient graph clustering algorithm and its application to natural language processing problems. In: Proceedings of the First Workshop on Graph Based Methods for Natural Language Processing, pp. 73–80. Association for Computational Linguistics (2006)
3. Damerau, F.J.: A technique for computer detection and correction of spelling errors. Commun. ACM **7**(3), 171–176 (1964)
4. FindTheData: Compare common words translated. http://common-words-translated.findthedata.com
5. Heeringa, W., Kleiweg, P., Gooskens, C., Nerbonne, J.: Evaluation of string distance algorithms for dialectology. In: Proceedings of the Workshop on Linguistic Distances, pp. 51–62. Association for Computational Linguistics (2006)
6. Johnson, R.A., Wichern, D.W.: Applied Multivariate Statistical Analysis, 6th edn. Pearson Prentice Hall (2007)
7. Kondrak, G.: N-gram similarity and distance. In: String processing and information retrieval, pp. 115–126. Springer (2005)
8. Levenshtein, V.: Binary codes capable of correcting deletions, insertions, and reversals. Sov. Phys. Dokl. **10**(8), 707–710 (1965)
9. Lin, D.: Automatic retrieval and clustering of similar words. In: Proceedings of the 36th Annual Meeting of the Association for Computational Linguistics and 17th International Conference on Computational Linguistics, vol. 2, pp. 768–774. Association for Computational Linguistics (1998)
10. Manning, C.D., Schütze, H.: Foundations of Statistical Natural Language Processing. MIT press (1999)
11. Pereira, F., Tishby, N., Lee, L.: Distributional clustering of English words. In: Proceedings of the 31st Annual Meeting on Association for Computational Linguistics, pp. 183–190. Association for Computational Linguistics (1993)
12. Philips, L.: Hanging on the metaphone. Comput. Lang. **7**(12), 39–43 (1990)
13. Serva, M., Petroni, F.: Indo-European languages tree by Levenshtein distance. EPL (Europhys. Lett.) **81**(6), 68,005 (2008)
14. SIL International: Ethnologue 17th edition website. http://www.ethnologue.com/ethno_docs/introduction.asp
15. Swadesh, M.: Salish internal relationships. Int. J. Am. Linguist. **16**, 157–167 (1950)
16. Swadesh, M.: The origin and diversification of language. Transaction Publishers (1971)
17. Wikipedia: Swadesh list. http://en.wikipedia.org/wiki/Swadesh_list
18. Xu, R., Wunsch, D., et al.: Survey of clustering algorithms. IEEE Trans. Neural Netw. **16**(3), 645–678 (2005)

The Design and Partial Implementation of the Dementia-Aid Monitoring System Based on Sensor Network and Cloud Computing Platform

Dingkun Li, Hyun Woo Park, Minghao Piao and Keun Ho Ryu

Abstract Sensor networks integrated with cloud computing platform provides a promising way to develop monitoring system for elderly people especially dementia patients who need particular care for their normal life. In our work, we aim to design a comprehensive, unobtrusive, real-time, low-cost but effective monitoring system to help caregivers for their daily healthcare work. The system design has been finished and the entire experimental environment including sensor network and cloud model has been set up in our lab to collect simulated data and one health care center to collect real data. Though the system has been partially implemented due to time limit, the experiment results are encouraging.

Keywords Healthcare monitoring · Sensor network · Cloud computing · Data mining

1 Introduction

The age span of the human beings has been greatly increased during the last 50 years, the rate of the dementia patient is increasing as well. Projections show that between 2000 and 2050 the number of people above the age of 60 will increase from 11 to 22 % worldwide, meaning that persons in this age group will number

D. Li · H.W. Park · K.H. Ryu (✉)
Database/Bioinformatics Lab, School of Electrical and Computer Engineering, Chungbuk National University, Cheongju, South Korea
e-mail: khryu@dblab.chungbuk.ac.kr

D. Li
e-mail: jerryli@dblab.chungbuk.ac.kr

H.W. Park
e-mail: hwpark@dblab.chungbuk.ac.kr

M. Piao
Department of Computer Engineering, Dongguk University Gyeongju Campus, Gyeongju, South Korea
e-mail: myunghopark@gmail.com

R. Lee (ed.), *Applied Computing & Information Technology*, Studies in Computational Intelligence 619, DOI 10.1007/978-3-319-26396-0_7

approximately 2 billion [1]. The healthcare for dementia patients among different counties is becoming a very important research issue.

In recent years, significant progress has been achieved in delivering health-related systems which is designed to achieve goal to improve dementia health care as well as elder people. Sensor network (SN) technologies have the potential to change the way of living with many applications in medicine care of the dependent people, and emergency management and many other areas [2]. In the past 10 years, cloud computing develops very quickly and provides a new way to establish new health care system in a short time with low cost.

A big amount of the researches have been done so far for older people's healthcare such as dementia [3, 4]. Applications that focus on fall and movement detection focus on following user movements and detecting falls. One example of this kind of service is Smart HCN, which consists of a WSN that monitors the posture of the subject and images taken by cameras to alert a specialist if the subject has suffered an accident [5]. Location tracking applications are based mainly on the principle of identifying the location of users and analyzing their behavior. One example of this kind of application is presented in work done by Lopez-Nores et al. [6]. An intelligent home monitoring unit has been developed to record the basic home activities and respond immediately when there is a change in the regular daily activity of the elder person [7]. In [8], it developed a non-invasive sleep monitoring system to distinguish sleep disturbances based on multiple sensors. All these works provide good examples for our work but some of them are obtrusive, expensive or hard to deploy. We aim to develop a comprehensive unobtrusive, real-time, low-cost but effective monitoring system for elder people especially dementia patients who need particular care for their normal life.

This paper comes from a real project of developing a dementia-aid monitoring system of Korean national project. This system mainly focused scenarios include fall over detection, getting out of the safety area detection, daily activities pattern analysis, and sleeping monitoring that cover most part of the patient health care.

The contributions of our work are: (1) designed a comprehensive monitoring system which covers main aspects of the health care. (2) The system is designed according to the unobtrusive, easy to deploy and effective principles. (3) It integrates manual records and sensor records, thus guarantee the overall precise. (4) Allowing applications to seamlessly handle sensor nodes and service mobility, and provides flexible communication between system and users. (5) Minimum the cost by using cloud computing platform which also guarantees real-time data transaction.

The rest of the paper is organized as follows: We describe the related work in Sect. 2. An overview of our system will be introduced in Sect. 3 and its implementation detail will be described in Sect. 4 followed by experiment result in Sect. 5. Section 6 concludes our work and depicts future work.

2 Related Work

This section briefly describes the sensor network, cloud platform and some key techniques we used in our work.

2.1 *Sensor Network (SN)*

Sensor network is dense wireless or wired network of amount of the sensor nodes, which collect and disseminate environmental data. In healthcare area, sensor network can be divided into two types, Body Area Network (BAN) and Personal Area Network [2]. BAN includes one or more types of sensors carried by the patient, while PAN includes one or more types of sensors deployed in the environment. In our system we avoid to use BAN because it will affect patient's normal life in most cases that obeys our unobtrusive design principle. And most sensors we used are wireless.

2.2 *GCM and GCSql*

Google Cloud Messaging (GCM) for Android is a service that allows you to send data from your server to your users' Android-powered device, and also to receive messages from devices on the same connection. The GCM service handles all aspects of queueing of messages and delivery to the target Android application running on the target device, and it is completely free [9]. The architecture of GCM is shown in Fig. 1.

First android device sends device id, application id to GCM server for registration, upon successful registration GCM server issues registration id to android device. After receiving registration id the device will send it to third-party server (TPS). The server will store registration id in the DB for later usage.

Fig. 1 GCM architecture

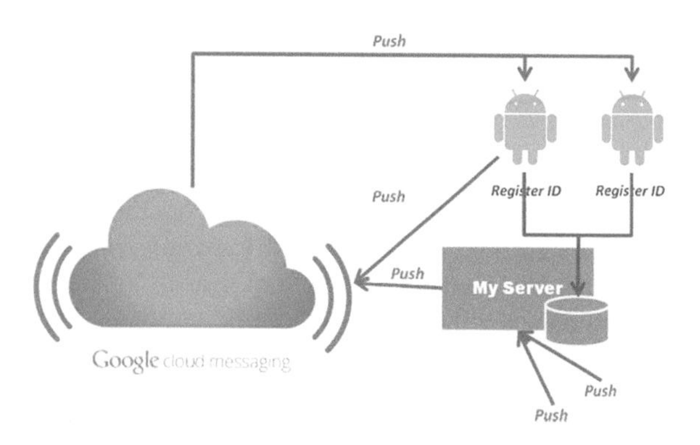

Google Cloud SQL (GCSql) uses MySQL deployed in the cloud and therefore the user gets all the benefits of using Explore Analytics with MySQL. Explore Analytics provides direct connectivity to Google Cloud SQL for live reporting, allowing you to deliver superb data analysis, visualization, and reporting. The data resides in Google Cloud SQL instance and there's no need to transfer the data to Explore Analytics [10].

2.3 *Data Stream Analysis and Time Series Analysis*

A data stream is an ordered sequence of instances that in many applications of data stream mining can be read only once or a small number of times using limited computing and storage capabilities [8]. Data Stream Mining is the process of extracting knowledge structures from continuous, rapid data records. Tuple is the basic element of the stream for processing. Two basic problems are the main issues for the streaming data are data sampling and filtering.

Since we cannot store the entire stream, we can store a fixed proportion or a certain period time of data. Two basic issues for stream data are sampling data, and filtering data. The Bloom Filter [11] strategy is a famous algorithm to do filtering.

A time series is a set of observations xt, each one being recorded at a specific time [12]. In our work most of the sensor data comes in the form of time series such as daily activities data, vibration data etc.

Some algorithms are very famous for dealing with time series data such as Dynamin Time Warping (DTW) [13], which is an algorithm for measuring similarity between two temporal sequences which may vary in time or speed. An improved version called FastDTW [14] is also widely used. Another algorithm used in our research is Longest Common Subsequence (LCSS) which is the problem of finding the longest subsequence common to all sequences in a set of sequences (often just two sequences) [15].

3 Main Framework

We give an overview of the whole system in this section, then we will describe the implementation detail in Sect. 4. The overview of the whole system architecture is depicted in Fig. 2.

(1) First the SN sends data streams (automatically) to the TPS which has been implemented by our team. Meanwhile we combine the manual record data such as activities records from patient, his relatives and nurse into our system to enhance the quality of the data source. After receiving the data, the TPS samples, filters, stores and transforms this data into appropriate form such as time series data. Some algorithms like DTW, LCSS, Bloom Filter etc. have

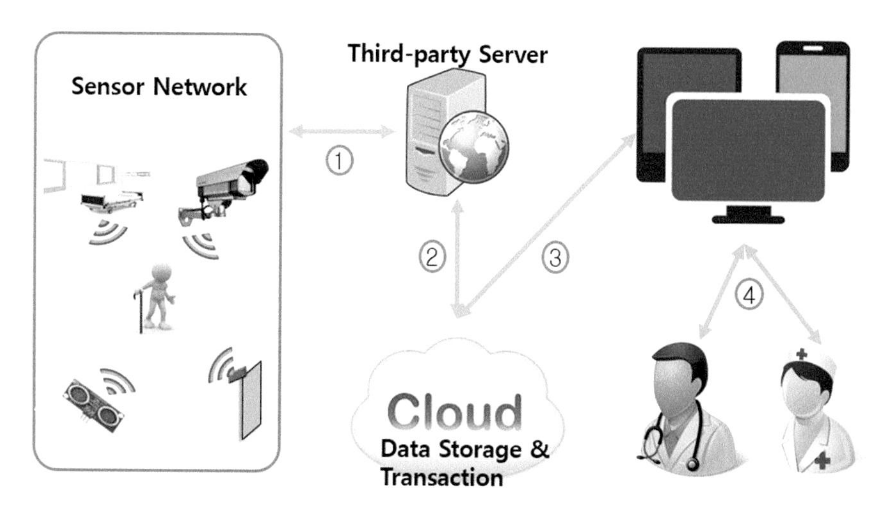

Fig. 2 System architecture

been used to process different kind of data to get the result for various scenarios.

(2) After processing step of TPS, it sends user request and processed data to the cloud model, which is used to store data and transfer data. This model is implemented by using GCSql and GCM cloud services.

(3) When receiving the requests from the TPS, cloud model responses immediately according to these requests, stores data or sends data to the devices registered to it.

(4) Finally, the end user, caregiver who interacts with the mobile devices to check the latest status of the patients. When there is an urgent situation such as fall detected by the TPS, it will give warning to the caregiver on the mobile devices through cloud model.

The system will be used for four scenarios which are show in Fig. 3.

The system divides the four scenarios into two categories, security and abnormal situation. When there is an emergency happened to the patient, such as falling over or walking out of the safe area, the system will give the warning to the caregiver

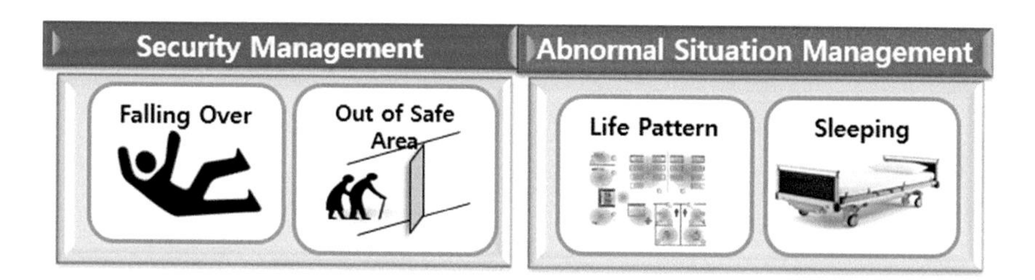

Fig. 3 Scenarios of the system

immediately. Also the system will analyze the normal life of the patient such as the daily activities and sleeping behaviors and after long term monitoring and analysis the system will distinguish the normal and abnormal situation (or pattern) of the patient and give warning to the caregiver when abnormal situation happened to the patient.

4 System Design and Implementation

The implementation detail of the whole system will be described in this section. Figure 4 shows the data flow of the system.

First, the system collects data by using SN, then the data will be transferred and processed by TPS. After that, the processed data will be stored in DB server and the common used and important data will be sent to and stored in the cloud model. More detail will be given as below.

4.1 Data Collection

For different scenarios, different sensors have been used to collect data we need. Table 1 lists the information of sensors.

All these sensors establish a SN which provides an all-round monitoring environment and sends data streams continuously for analysis. Furthermore, we tried our best to avoid the obtrusiveness and power supply problems. The sensors we used should not disturb the user normal activities especially in BAN and have an

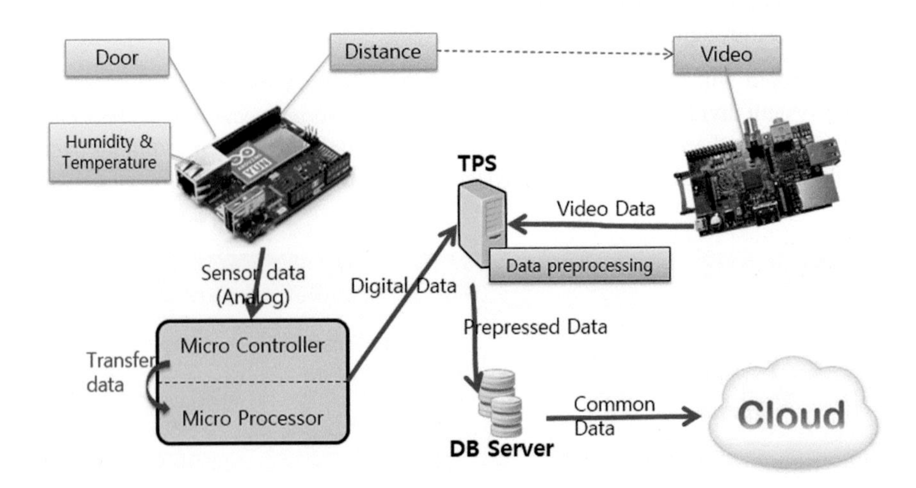

Fig. 4 Data flow of the system

Table 1 Sensor type we used for the system

Sensor type	Comments	Wireless
Ultrasonic sensor	Detect the object is inside the room or not	Y
Temperature and humidity sensor	Detect the temperature and humidity of the patient room	Y
Arduino	An inexpensive, adaptable, and programmable open source microprocessor that can read data input in the form of voltage at its analog pins	N
Raspberry Pi B+	Control camera	N
Door sensor	Detect the open of the door	Y
Vibration sensor	Detect the vibration signal	Y
RFID reader/writer	Deal with the RFID signal	Y
RFID TAG (900 MHZ)->Beacon	Wear on patient's wrist to send RFID signal to identify different patient	Y
Weight sensor	Fixed on the patient's bed to detect the sleeping time and posture	Y
Camera	Take pics without interference in privacy	N
Distance sensor	Detect the distance between patient and sensor	Y
Motion perception sensor	Detect the movement of the patient	Y

appropriate way for power supply. For example, avoid using ECG sensor which is usually too obtrusive to carry. The SN signal should cover the whole health care center to collect real data and most sensors are fixed.

4.2 Data Preprocessing

For the sensor network, we treat each sequence record sent by the sensor as a stream, so at the same time, the receiver may receive several streams. Furthermore, it is a continuous and unbounded procedure, as the time passing by, there could exist no big enough space to store all the stream data, and a lot of the data is not related to the system. So we need sampling method to store fraction of the data stream and filtering method to eliminate unrelated data stream.

For example we only choose three months data as sample that is 40 % fraction of the total we recorded so far, which is thought to be enough for analysis since one treatment for a patient is no more than 3 months in most cases.

Usually there are more than one person such as several patients use one room. The sensors will receive different persons' data streams since most of the sensors are fixed. So how to filter the specified data streams from the amount streams is becoming a practical issue. The Bloom Filter algorithm has been used to solve this problem. Figure 5 shows the detail.

- ❖ Input: A list of the patient properties value S such as weight, time series data of vibration, personal sensor id etc., which can distinguish the patient from others. An array B with its size n which is the appropriate size of the memory.
- ❖ Algorithm detail:

  Consider: **|S|** = m, **|B|** = n

  Use k independent hash functions $h_1,\ldots, h_k$

  Initialization:

  Set **B** to all **0s**

  Hash each element $s \in S$ using each hash function h_i, set **B**[$h_i(s)$] = **1** (for each $i = 1,\ldots, k$)

  Run-time:

  When a stream element with key x arrives

  If **B**[$h_i(x)$] = **1** <u>for all</u> i = **1,…,** k then declare that x is in S

  That is, x hashes to a bucket set to **1** for every hash function $h_i(x)$

  Otherwise discard the element x
- ❖ Output: Patient related data streams only

Fig. 5 Bloom filter algorithm

Another issue is data transformation. Some data such as patient's daily activities data, like "wake up, have breakfast, watch TV, do sports etc." got from his relatives is of string form. The system abstracts these activities and assign different id to them. By using this way the string data can be transformed into numerical form data very easily.

4.3 Data Storage and Transaction Through Cloud Model

We implemented the cloud model by following the APIs provided by google cloud platform. Cloud model plays the role of data storage and transfer and consists of public cloud services GCSql and GCM. Consider about the efficiency, connection and security problems, we only store common used, important and urgent information in GCSql database, such as falling over data, delivering data etc. While GCM is the key part for the communication between TPS and devices.

The most important feature of GCM in our project is that it delivers real-time data. When there is an urgent situation happened to the patient, the caregiver will receive notice send by TPS without much delay (usually within 1 s) through GCM, but the users need to keep the devices online all the time.

4.4 Implementation Detail

This subsection describes design and implementation details.

4.4.1 Falling Over

In our project, we detect patient falling over situation by using RFID transfer, distance sensor, vibration sensor and camera (camera will not be fixed in the restroom). The usage of the sensors is shown in Fig. 6.

The door sensor will be triggered when the patient opening the door, distance and vibration sensors will start to work to capture the status of the patient and send the data to the TPS. When there is a sudden change of the data received, the server treats this situation as a falling over situation and sends warning to the devices through cloud model of the system.

If the urgency is not detected inside the restroom such as living room, related camera will take photos of the current scene and send pictures to the server. Also these pictures will be stored in GCSql database.

When the devices received the warning data, they alert to the caregivers by giving notification, pop up window as well as warning beep. The caregivers check the devices and make the corresponding processing.

4.4.2 Walking Out of the Security Zone

In the health care center, patients usually have their safety area. Something unpredictable may happen to the patients if they leave this area without noticing the safe guard or caregivers. RFID transfer, motion perception sensor and camera have been used for this scenario. The usage of them is shown in Fig. 7.

Door sensor will be triggered if patient is opening the door, motion perception sensor is used to detect the direction of the patient movement, RFID Transfer detects which patient is walking out and camera is used to record the whole procedure.

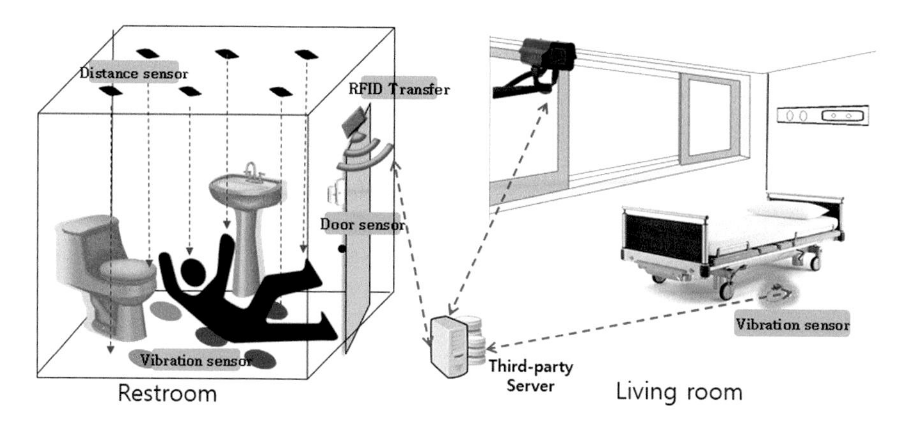

Fig. 6 Sensors deployment for falling over detection

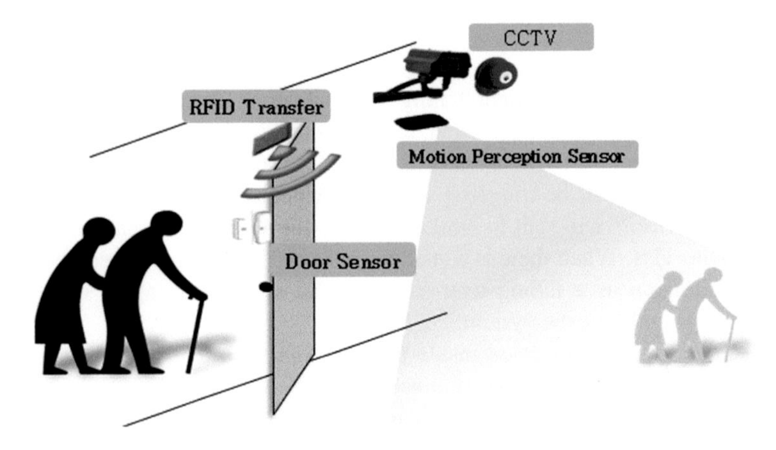

Fig. 7 Sensors deployment for walking out of the security zone detection

When patient is walking out of the safe area, this situation will be detected by the TPS, after analyzing the received data from sensors, it will send warning data to devices though cloud model. The caregivers check the devices and make the corresponding processing.

4.4.3 Abnormal Life Pattern Detection

We define abnormal Life Pattern as the activities which are very different from the other days. For example, go to the toilet too many times in one day.

A significant amount of research about daily activities detection has been already done and is still undergoing such as [7], there is still a lack in the confirmation of a suitable technique and a low-cost solution.

We provide an easy, low-cost but efficient way to detect patient life pattern. Door sensor, RFID transfer, motion perception sensor etc. have been used with their usage shown in Fig. 8.

After assigning each activity an id, the data processed by TPS is of time series data form. We treat one day data as one time series, so after a certain period such as one quarter, we can have approximate 100 time series records, and these records are used as the training set. The next step is show in Fig. 9.

We use DTW algorithm to calculate the similarity between each time series record. Then the KNN algorithm will be used to classify these records according to the similarities we get. After this step we will have k classes. An assumption has been made in this situation: the class contains the most elements is the normal daily activity class N, because for a normal person the routine is more or less the same for a certain period.

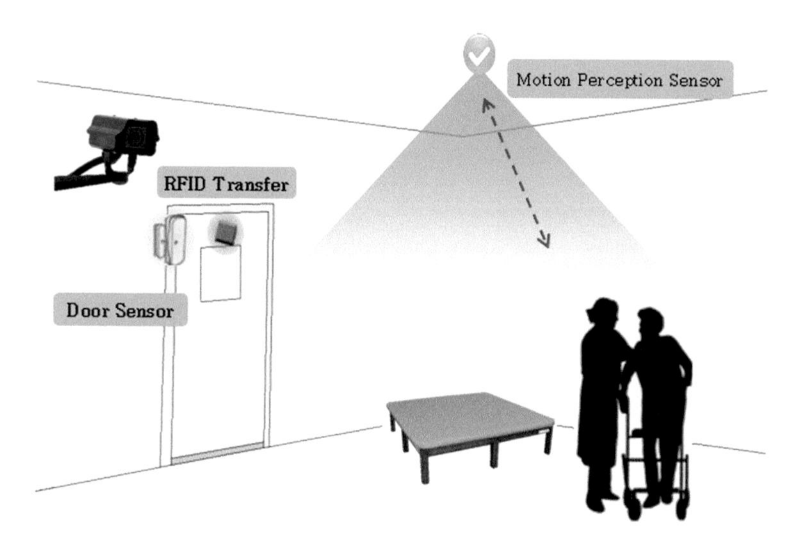

Fig. 8 Sensors deployment for abnormal life pattern detection

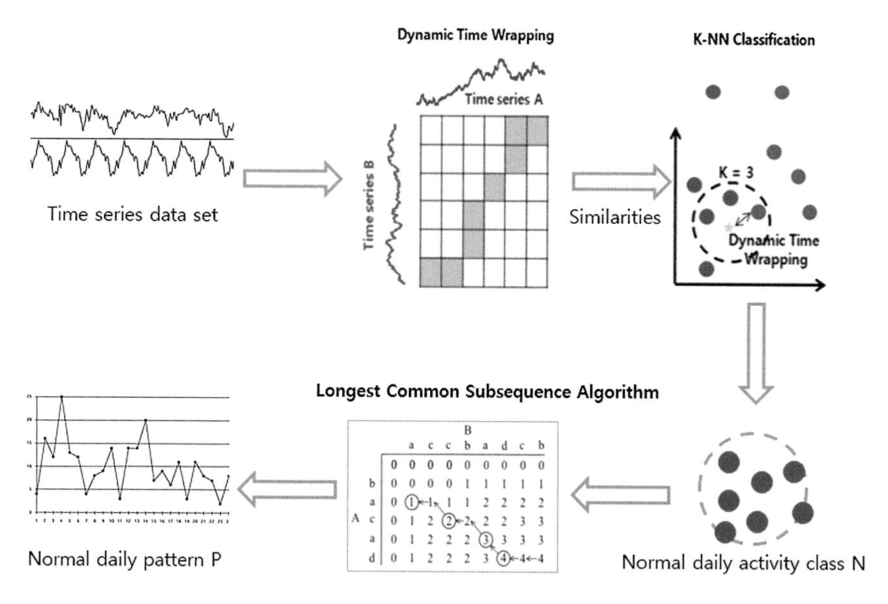

Fig. 9 Normal daily life pattern mining procedure

By using the LCSS algorithm on N we can abstract a daily life pattern P that is the most common daily activities. The daily activities data collected from the relatives is used to amend the life pattern we generated by using LCSS algorithm.

When new record has been received after one day monitoring, the TPS transforms it to time series record T and DTW algorithm will be used to calculate the

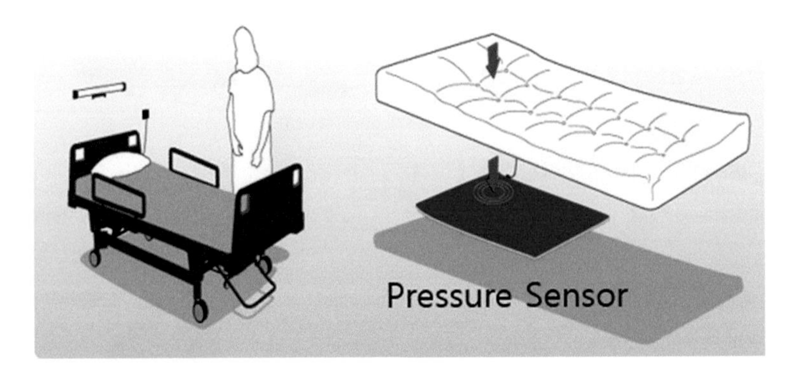

Fig. 10 Pressure sensor usage for sleeping behavior detection

similarity between P and T. If the similarity is smaller than a user predefined threshold, TPS will treat this situation as abnormal life pattern. When TPS detected this pattern, it continuously to find and analyze the activity or activities that caused this abnormal pattern. After that it will send warning to the devices through cloud model. The caregivers check the devices and make the corresponding processing.

4.4.4 Sleeping

The sleep quality, such as time, duration and frequency, is one of the most important indicators to patient health. We designed an easy way to monitor patient sleeping habit. In our system, a pressure sensor has been used to collect data of the patient sleeping behavior. It is fixed on the bed to record start time and end time of each sleep. The pressure data is also recorded to help distinguish different people. Figure 10 shows the overlook for this scenario.

By analyzing the start time and end time for each sleep, we can calculate the time, duration and frequency of the patient sleep thus evaluate the quality of the sleep.

We analyze long-term sleeping habit of the patient, do mining work on the data collected to find a normal sleeping pattern. By comparing this pattern with new coming sleeping data the system can find irregular or abnormal sleeping situation therefore notice the caregiver to handle it.

5 Experiment

Recently, we have finished the design of the whole system, but the implementation is still undergoing. Some models of the system have been finished. For example, the SN has been established in our lab (some sensors are shown in Fig. 11) and some sensors have been used in health care center to collect real data. We combined

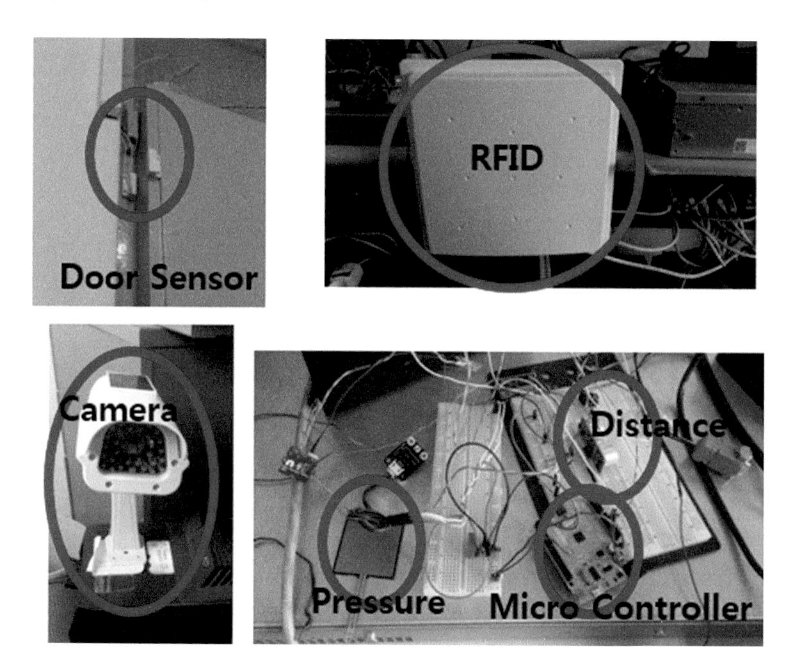

Fig. 11 Sensors deployed in the lab

simulated data and small number of real data as test data for system implementation. We use MySQL DBMS as the data storage center of TPS, and some algorithms of TPS such as DTW, KNN which are modified to fit for our data type and user requirements have been implemented. We have implemented the cloud model by following the APIs provided by google cloud platform. Right now the TPS can interact with the android devices through cloud model. Several devices have been used for the purpose of testing include Nexus 5, Nexus 7, LG G pad 8.3, etc.

One experiment we did is abnormal life pattern detection. Figure 12 shows the data we collected for daily activities, and the TPS analysis is shown in Fig. 13.

Start time	End time	Location	Type	Place
2014-11-11 21:14:21	2014-11-12 00:21:49	Seat	Pressure	Living
2014-11-12 00:22:57	2014-11-12 00:22:59	Door	PIR	Living
2014-11-12 00:23:14	2014-11-12 00:23:17	Door	PIR	Kitchen
2014-11-12 00:24:20	2014-11-12 00:24:22	Door	PIR	Kitchen
2014-11-12 00:24:42	2014-11-12 00:24:54	Door	PIR	Living
2014-11-12 00:25:35	2014-11-12 00:42:56	Seat	Pressure	Living
2014-11-12 00:43:46	2014-11-12 00:43:49	Door	PIR	Living
2014-11-12 00:46:12	2014-11-12 00:46:15	Door	PIR	Bedroom
2014-11-12 00:47:21	2014-11-12 00:47:24	Door	PIR	Bedroom
2014-11-12 00:48:38	2014-11-12 00:50:12	Basin	PIR	Bathroom
2014-11-12 00:50:29	2014-11-12 00:50:32	Door	PIR	Living

Fig. 12 Record data format of the daily activities

(Unit: hour)

Name	Indoor Activities	Ourdoor activities	Treatment	Sleep	Toilet	Shower	Others
김상O	14	10	1	6	1	1	5
박민O	16	8	1	6.5	1	1	4
홍성O	18	6	2	7	1	0	1

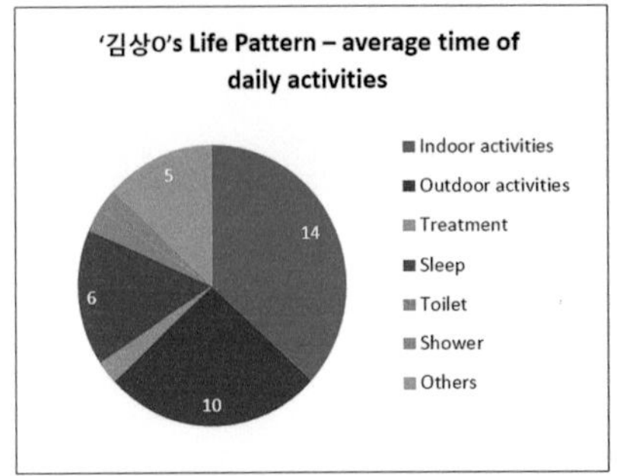

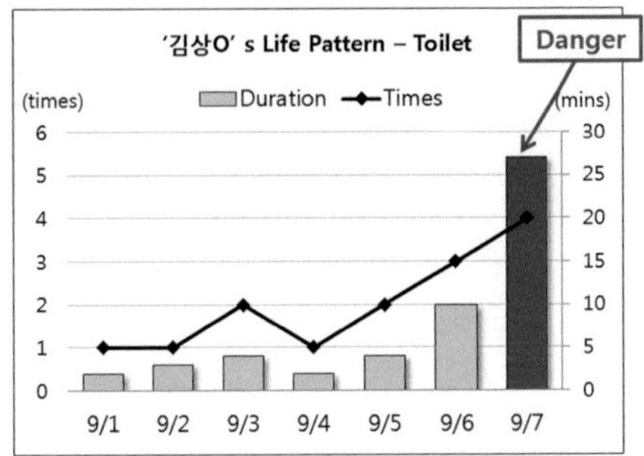

Fig. 13 Abnormal pattern analysis result by TPS

Fig. 14 **a** Warning received from the cloud model. **b** Falling over picture captured by camera

When the TPS detected an abnormal life pattern, it continues to analyze the particular abnormal activity or activities that causing this abnormal life pattern. For example going to the toilet too many times marked in red shown in Fig. 13 is detected by the TPS, and it will sent warning to the user devices.

Another experiment we did was falling over detection. Simulation data was used in this case. When this situation had been detected by TPS, it sent warning to the devices immediately. The result is show in Fig. 14a. And the Fig. 14b shows the falling over picture captured by camera.

6 Conclusion and Future Work

The cloud computing has greatly changed the way of developing software in many ways, it provides a fast, low-cost, safe and easy way to design new software. In our work, we designed and partially implemented a comprehensive monitoring system which consists of four models: SN, TPS, cloud and user interface models for dementia patients based on the sensor network and cloud computing. The system is designed and developed according to the unobtrusive, easy to deploy, effective, low-cost and real-time principles.

For the next stage of research, first thing is to implement the whole system, then do in-depth simulations to validate the system performance, in particular, simulate multiple scenarios to confirm its scalability so as to apply our system to real environment. Some algorithms are under the consideration to improve their performance such as using FastDTW to replace DTW, using slide window to improve the Bloom Filter algorithm. We are considering to cover more scenarios, such as medical status monitoring, medication intake, etc. We also plan to extend our system to iOS platform in future.

Acknowledgements This work was supported by the National Research Foundation of Korea (NRF) grant funded by Basic Science Research Program through the National Research Foundation of Korea (NRF) (No. 2013R1A2A2A01068923) and the MSIP (Ministry of Science, ICT and Future Planning), Korea, under the ITRC (Information Technology Research Center) support program (NIPA-2013-H0301-13-4009) supervised by the NIPA (National IT Industry Promotion Agency).

References

1. WHO Website, Interesting Facts about Ageing, Available online: http://www.who.int/ageing/about/facts/en/index.html. Accessed on 20 Mar 2015
2. Alemdar, H., Ersoy, C.: Wireless sensor networks for healthcare: a survey. Comput. Netw. **54**(15), 2688–2710 (2010)
3. Chung-Chih, L., Ming-Jang, C., Chun-Chieh, H., Ren-Guey, L., Yuh-Show, T.: Wireless health care service system for elderly with dementia. IEEE Trans. Inf. Technol. Biomed. **10**(4), 696–704 (2006)
4. Chung-Chih, L., Ping-Yeh, L., Po-Kuan, L., Guan-Yu, H., Wei-Lun, L., Ren-Guey, L.: A healthcare integration system for disease assessment and safety monitoring of dementia patients. IEEE Trans. Inf. Technol. Biomed. **12**(5), 579–586 (2008)
5. Tabar, A.M., Keshavarz, A., Aghajan, H.: Smart home care network using sensor fusion and distributed vision-based reasoning. In: Proceedings of the 4th ACM International Workshop on Video Surveillance and Sensor Networks, pp. 145–154. Santa Barbara, CA, USA, ACM, Santa Barbara, USA, 27 Oct 2006
6. Lopez-Nores, M., Pazos-arias, J., Garcia-Duque, J., Blanco-Fernandez, Y.: Monitoring medicine intake in the networked home: the icabinet solution. In: Proceedings of the Second International Conference on Pervasive Computing Technologies for Healthcare (PervasiveHealth 2008), pp. 116–117. Tampere, Finland (2008)

7. Suryadevara, N.K., Gaddam, A., Rayudu, R.K., et al.: Wireless sensors network based safe home to care elderly people: behaviour detection. Sens. Actuators A Phys. 277–283 (2012)
8. Wolford, D.K.: System, pad and method for monitoring a sleeping person to detect an apnea state condition: U.S. Patent Application 12/359,459[P] (2009)
9. https://developer.android.com/google/gcm/gcm.html#arch. Accessed 2 Jan 2015
10. https://cloud.google.com/sql/. Accessed 5 Feb 2015
11. Bloom, Burton H.: Space/time trade-offs in hash coding with allowable errors. Commun. ACM **13**, 422–426 (1970)
12. Chatfield, C.: The analysis of time series: an introduction. CRC press (2013)
13. Sakoe, H., Chiba, S.: Dynamic programming algorithm optimization for spoken word recognition. IEEE Trans. Acoust. Speech Signal Process. ASSP **26**, 43–49 (1978)
14. Stan, S., Philip, C.: FastDTW: Toward accurate dynamic time warping in linear time and space. In: KDD Workshop on Mining Temporal and Sequential Data, pp. 70–80 (2004)
15. Bergroth, L., Hakonen, H., Raita, T.: A survey of longest common subsequence algorithms. In: Proceedings of Seventh International Symposium on String Processing and Information Retrieval, 2000. SPIRE 2000, pp. 39–48. IEEE (2000)

Experimental Use of Learning Environment by Posing Problem for Learning Disability

Sho Yamamoto, Tsukasa Hirashima and Akio Ogihara

Abstract In this paper, we described about a design of learning environment based on information structure and a realization of problem posing for learning disability. We design and developed software by modeling information structure of subject that is operated on software. In this research, we aimed at the domain of education and developed a learning environment for posing arithmetic word problem. Problem posing is well-known as effective learning. But students who have learning disability are impossible to learn by this exercise because they cannot pose the problem from scratch and cannot read complex sentence. However, learner is able to pose a word problem by selecting three cards in our software. So we assumed that these learners are able to learn by problem posing by using our learning environment. Experimental use is also reported.

Keywords Tablet PC · Multimedia · Interactive · Learning disability · Problem posing

S. Yamamoto (✉) · A. Ogihara
Department of Informatics, Faculty of Engineering, Kinki University,
1 Takaya Umenobe, Higashi-Hiroshima City, Hiroshima 739-2116, Japan
e-mail: yamamoto@hiro.kindai.ac.jp

A. Ogihara
e-mail: ogihara@hiro.kindai.ac.jp

T. Hirashima
Graduate School of Engineering, Hiroshima University, 1-4-1 Kagamiyama,
Higashi-Hiroshima City, Hiroshima 739-8527, Japan
e-mail: tsukasa@lel.hiroshima-u.ac.jp

R. Lee (ed.), *Applied Computing & Information Technology*,
Studies in Computational Intelligence 619,
DOI 10.1007/978-3-319-26396-0_8

1 Introduction

Recently, ICT (Information and Communication Technology) is widespread our life and many people easy to use these technologies. We design and develop software based on ICT. Our software is designed by modeling a subject that is operated on it. Now, our target is Education. In domain of education, teacher is easy to manage a course for learning in their lesson by using Moodle [1]. MOOC (Massive open online course) realized that a learner is able to take various lessons via the web [2]. On the other hands, a several researchers are working on a realization of difficult learning by ICT.

For example, some researchers suggested that the realization of problem posing that is very difficult exercise without software [3, 4] because a teacher has to assess a posed problem by posing their students and a student poses a problem from scratch. The learner is required to pose a multiple-choice question from scratch and they assess problems by posed other learners in their software. Here, in our research, we have redesigned the problem posing exercise and developed software for the problem posing [5]. In this learning environment, a learner is required to pose a problem by selecting three cards and arranging them in proper order. So, this learning environment is able to perform the assessment of posed problem automatically.

It is known that problem posing is very difficult exercise for a learner who has learning disability like language delay because these learners cannot pose the problem from scratch because of their disability. But we assumed that our learning environment is suitable for these learners to pose problem except for their disability because we redesigned the problem posing exercise based on information structure of the operated subject on software. So, our purpose in this research is to confirm that a learner who has learning disability is able to exercise the problem posing by using our learning environment. In this paper, we described about a comparison of a design of our software with a characteristic of language delay. An experimental use of our learning environment is also reported.

2 Problem Posing and Learning Disability

2.1 Problem Posing

Several researchers have suggested that problem posing is effective method for learner to master the solution method [6, 7]. In this exercise, a learner is required to pose a problem based on a given calculation, given situation and so on. Figure 1 shows an example of arithmetic word problem that is targeted in this research. This word problem can be solved by one-step addition or subtraction. The learner is required to pose a problem like this example, he/she is given an assessment that pose the problem that can be solved by "$6 - 2 = \square$" ($\square$ means required value).

There is one big tree. Tree has six apples and there are five oranges on other tree. Then, two apples are picked. How many apples are there?

Fig. 1 Example of arithmetic word problem

Then, the learner includes the several modifying phrases. For example, "there is one big tree". He/she also includes non-related information that is "there are five oranges on other tree". And, a posed problem may be included some syntax errors if the learner don't enough to learn a language. Therefore, it is hard for teacher to assess a posed problem. Moreover, there are from thirty to forty students in class. So, problem posing is difficult exercise by realizing in class.

2.2 Targeted Learning Disability

Target group in this research is a learner who has learning disability (LD). Generally, these learners learn an arithmetic word problem more slowly and carefully than a learner who is not has LD [8, 9]. So, these learners are able to solve easy problems but it is very difficult to exercise by problem posing. And, several learners who have language delay cannot exercise the problem posing. Language delay means that children are fail to develop their language abilities appropriate for their age. One of them cannot pose a sentence from scratch and difficult to read a complex sentence. For example, one student can read "there are five apples" but cannot read "There are five apples but Tom eat two apples". A teacher who has performed the experimental use in this paper said that his students are not able to pose a problem because of this disability.

3 Design of Our Learning Environment

In our research, we have analyzed an information structure of arithmetic word problem because a learner who uses our software operates these word problems. First, we designed the information structure of targeted arithmetic word problems. Second, a problem posing by operating this information is designed.

3.1 Information Structure of Arithmetic Word Problem

Our target domain is one-step addition or subtraction word problem that learner learns in the 1st grade student of an elementary school. Example of this word problem is shown in Fig. 2. A learner is required to pose a problem like this in

There are six doughnuts and you eat two doughnuts for a snack. How many doughnuts are there?

Fig. 2 Example of arithmetic word problem

There are six doughnuts.
Two doughnuts are eaten.
There are several doughnuts.
How many doughnuts are there?

Fig. 3 Example of arithmetic word problem based on triplet structure model

problem posing exercise. In other words, the learner is required to operate the word problem in problem posing exercise.

This word problem consists of three values: two given values and one required value. And, these values mean quantity of given object. For example, "six" means a quantity of doughnuts. Then, "doughnut" is object. These three values have other property. "There are six doughnuts" and "How many doughnuts are there?" mean existence of object. "You eat two doughnuts" means relation between two existence objects. Then, "eat" means subtraction. We call these sentence as "existence sentence" and "relational sentence". These properties can be judged by predicate. For example, "eat". We call this model as triplet structure model [10].

Figure 3 shows the arithmetic word problem by modifying based on triplet structure model. Each sentence is consists of object, value and predicate. And this word problem consists of two existence sentences and one relational sentence. In this problem, "for a snack" is removed but this word is not used for solving this word problem. And first sentence is divided two sentences for the same reason.

Moreover, this arithmetic word problem is expressed by four stories. These are combine, change and compare. Story that is expressed a change is divided into increase and decrease. Figure 2 is the example of "decrease story". These stories are consists of two existence sentences and one relational sentence but an expression of relational sentence are different. For example, "There are nine A and B in total" means "combine story".

3.2 Comparison of Design of Problem Posing with Learning Disability

By triplet structure model, the arithmetic word problem that can be solved by one-step addition or subtraction is expressed by three sentences. These sentences consist of value, object and predicate. In this section, we explain that a design of problem posing based on this model and comparison of our problem posing with LD.

In our model, a learner is given an assignment that is contains a calculation and a story. For example, "Pose the problem that can be solved by 6 − 2= □ and change-increase story". Then, the learner is also given several sentence cards contain correct cards and dummy cards that is not necessary to pose a correct problem. Each card consists of object, value and predicate. He/she poses a problem by selecting three sentence cards and arranging them in proper order.

We have already described the characteristic of target LD in Sect. 2. It is very difficult for these learners to pose problem from scratch and to understand sentence that is combined more than two sentences. However, in our model, some word for modifying expression like "for a snack" and complex sentence that is combined more than two sentences should be removed. They are able to pose a problem by building given sentence cards. Therefore, learner who has LD is able to understand a meaning of word problem like Fig. 3. And then, he/she comes to be able to exercise by problem posing without their disability.

4 Learning Environment

4.1 Summary of Learning Environment

We explained about a summary of our learning environment for learning problem posing. This learning environment consists of two software and one server application. A software for exercising by problem posing called MONSAKUN Touch. This software is developed by JAVA and run on Android Tablet. Other software for monitoring a learning data on MONSAKUN Touch called MONSAKUN Monitoring. This software is developed by PHP and Javascript. So, MONSAKUN Monitoring run on browser software. These software are connected via database server that RDBMS is used MySQL. This server is built by Red Hat and web server application is used Apache 2.0.

MONSAKUN Touch send a result of learner's exercise to database server when they answer an assignment. MONSAKUN Monitoring show a teacher this result by referring a data on database in real-time.

4.2 MONSAKUN Touch

User interface for problem posing is displayed in Fig. 4. First, a learner logging in MONSAKUN Touch, and then he/she is required to select a level of assignment. This level has five steps. When the learner selected one level, this interface is shown. The left side is problem composition area. On the top of problem composition area, a number of level, a number of assignment and the sentence of assignment is given. The sentence of assignment is described a calculation and a

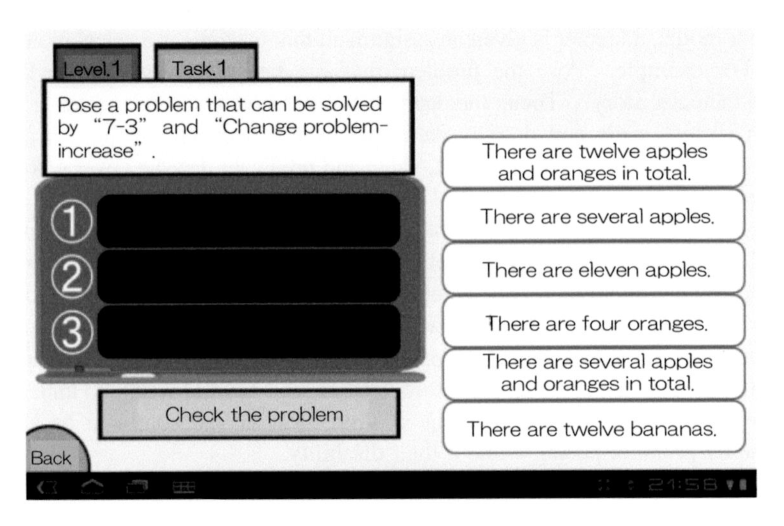

Fig. 4 User Interface for posing problem

story. On the right side, six sentence cards are given. The learner is able to move these sentence cards by dragging card with finger. These cards are consists of three correct cards and three incorrect cards. The learner is required to pose problem by selecting three sentence cards and arranging them in proper order. Then, he/she set the selected cards to three blank on the middle of left side. If the learner fills in all blank by sentence card, a button of "Check the problem" is changed a button of "Assess the problem". Then, if the learner taps this button, a posed problem is assessed by software automatically. When the learner answers all assignments, the selected level is finished and this interface is changed to interface of selected level.

Figure 5 shows a process of assessment. This software assesses a posed problem based on information structure of arithmetic word problem that we described in Sect. 2. First, MONSAKUN Touch assesses the structure of sentence. This means that the software check the posed problem that includes two existence sentences and one relational sentence. When it doesn't satisfy this constraint, this software feed back the learner "This problem is not include a story". Second, if posed problem is satisfied the structure of sentence, this software checks a combination of each object and each value. When the posed problem is not satisfied this constraint, this software feed back it. After that, if the posed problem consists of correct combination of objects and values, the correct problem was posed. But this problem may not satisfy an assessment that is given a calculation and a story. So, last, the software assesses the posed problem that corresponds with the assignment. Of cause, if the posed problem doesn't correspond with the assignment, this software feed back it. For this assessment, the software gives the learner the rich feedbacks automatically.

When the learner taps the button for assessment, these results of assessment send to database server. The contents of this data are a number of level, a number of assignment, each sentence of selected three cards, a result of assessment and

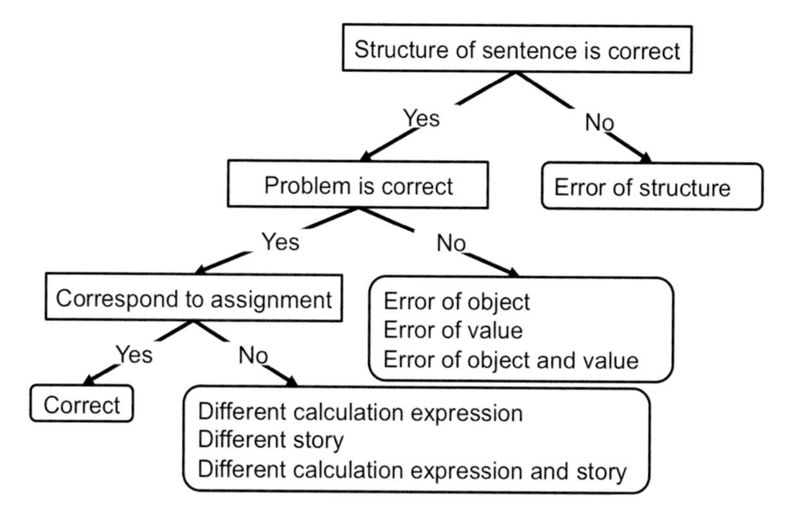

Fig. 5 Process of assignment

timestamp. Then, the result of assessment is send as number. For example, correct is zero and error of structure is one. And then, the software sends a number of correct problems, a number of incorrect problems and a number of each error in this time. We called these data as learning data.

4.3 MONSAKUN Monitoring

Figure 6 displays a user interface of MONSAKUN Monitoring. This software shows a teacher a result of exercise by getting the learning data from database server. Figure 6 shows the learning data in each classroom. One square means one lesson. The data of one lesson includes the bar chart, the doughnuts chart and some text. The bar chart means the each number of posed problems, correct problems and incorrect problems. The doughnuts chart means the rate of each error. For example, the error of structure, the error of object and so on. Each data is an average of this classroom. The text on top area shows the date of lesson. And the text below the date of lesson shows the average progress of exercise in this lesson. For example, "Level 1, Assignment #10".

If teacher click a link that is described "show student data", this software shows the each student's learning data in selected lesson. Then, teacher can check the log of problem posing in each student. By using this software, teacher is able to check the student's learning in real-time.

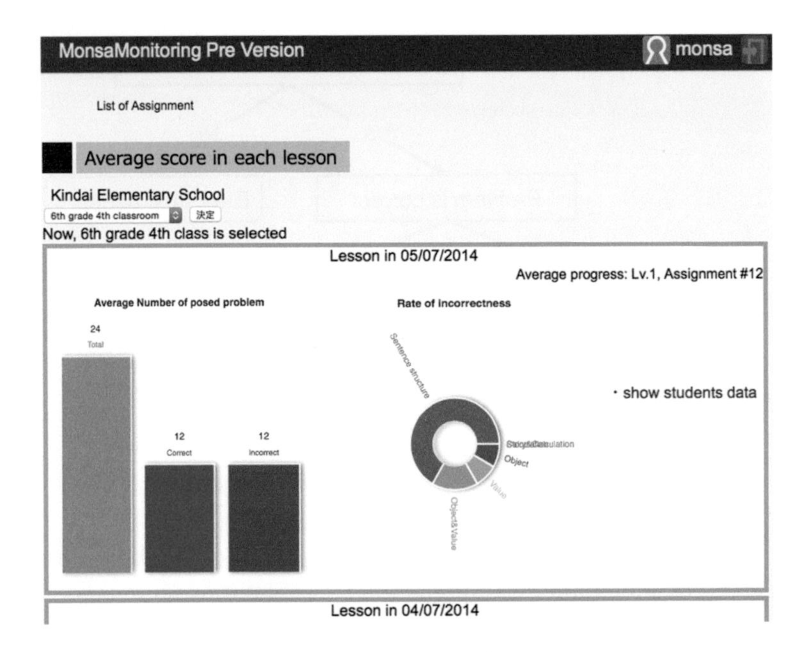

Fig. 6 User Interface of MONSAKUN Monitoring

5 Experimental Use

5.1 Procedure of This Experiment

The subjects are two students who belong a public school in Nagasaki. They learn in a special classroom. One student is 5th grade student and other one is 6th grade student. Both subjects are male. Next, We described the characteristic of these subjects. 6th grade student is able to calculate by one, two and three digit in the one-step addition or subtraction. On the other hands, 5th grade student is able to calculate by one and two digits but he cannot calculate by three digits. It is difficult for these subjects to understand a complex sentence. They can solve the arithmetic word problem for 1st and 2nd grade students. But, if the word problem has complex sentence, the word problem becomes very difficult for them. And, they cannot make a sentence from scratch. So, a homeroom teacher judges that they cannot pose the problem from scratch. Therefore, these subjects are suitable for using our software and investigating an effect of our software.

A procedure of this experiment is described. This experiment has used six lessons (one lesson is forty five minutes, three weeks and six days). These subjects have worked on a pretest and posttest before and after these lessons. Each lesson is consists of a teacher's teaching about a method of problem posing in 30 min and an exercise of MONSAKUN Touch in 15 min approximately. The teacher has taught one level in one lesson. So, subjects have learned all level from 1st to 5th lesson and 6th lessen used for review of level 1 in 1st lesson.

We explain about a system environment in this experimental use. We have provided the database server and our learning environment for this classroom. Network has been built by tethering of teacher's smart phone. We have used two android tablets with 10.1 inches monitor that are the property of the school. Teaching assistant for software is nothing.

5.2 *Material for Pretest and Posttest*

The subjects have carried out a usual problem-solving test and an extraneous problem-solving test for pretest and posttest. Usual problem-solving test is an arithmetic word problem that can be solved by one-step addition or subtraction. But these problems are described based on triplet structure model. This test was performed for measuring subject's problem-solving performance. Extraneous problem test includes extraneous information that is not necessary to solve the word problem. In this test, subjects can distinguish the problem structure that we described in Sect. 2. This test was performed for measuring a learner's understanding of problem structure [11]. There are sixteen problems in each test.

5.3 *Results and Discussion*

For considering about a result of this experimental use, we formed three hypotheses. (1) Subjects can pose several problems. (2) If subjects can pose some problem, their performance for posing problem is improved in an exercise of same level. (3) If subjects can pose the problem, they are able to improve their problem-solving performance and their understanding of problem structure.

First, hypothesis (1) is examined. The log of problem posing exercise on MONSAKUN Touch is shown in Fig. 7. A vertical axis means a number of posed problems and an accuracy rate of each exercise. A horizontal axis means a number of lesson and a level that is taught by teacher in each lesson. Average number of posed problem is twenty-three problems in each lesson and accuracy rate is 48 %. On the other hands, in the result of 1st grade students in regular class, average number of posed problem is twenty-two (about 8 min) and accuracy rate is 56 %. Accuracy rate in level 4 sharply decreased but an assignment of level 4 is difficult for a student without LD. Therefore, this result suggested that subjects are able to pose a problem by using MONSAKUN Touch.

Second, hypothesis (2) is considered. Because difficulty of each level is different, we compared 1st lesson and 6th lesson. In Fig. 7, the accuracy rate in 6th lesson is higher than in 1st lesson. And, a changing the rate of each error is shown in Figs. 8 and 9. This bar chart shows the number of each error in 1st lesson and 6th lesson.

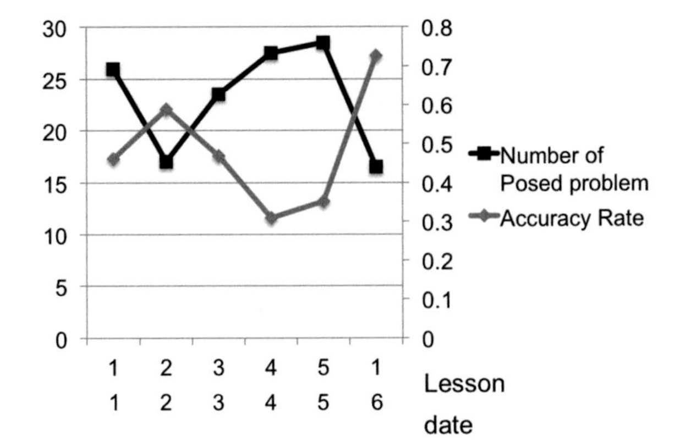

Fig. 7 Log of problem posing

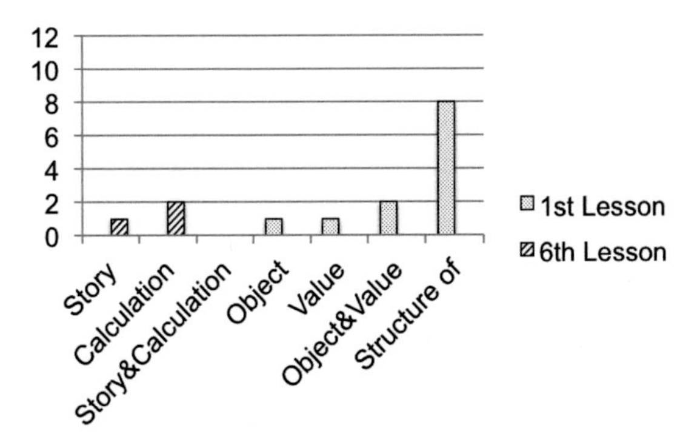

Fig. 8 Changing the rate of each error (6th grade student)

Most incorrectness in 1st lesson is error of structure in Fig. 8. This means the 6th grade student didn't understand the problem structure that consists of one relational sentence and two existence sentences in this exercise. However, in 6th lesson, this incorrectness is nothing. This result suggested that the 6th grade student was able to notice at least structure of sentence for posing a correct problem. On the other hands, in Fig. 9, there are various errors in the result of 5th grade student more than 6th grade student in 1st lesson. But most of error is the structure of sentence. This point is same as the result of 6th grade students. In 6th lesson, the kind and number of error is decreased but the structure of sentence is remained. This result suggested that both subjects improved their performance of problem posing but the result of learning is not sufficiently in 5th grade. Leaning disability about language delay has several degrees. So, the given sentence cards are suitable for the degree of 6th grade student's disability but not suitable for 5th grade. Therefore, we have to study about the degree of language delay and redesign the given information of this software.

Last, we described about hypothesis (3). The results of pretest and posttest are shown in Table 1. Because subjects are not many, we cannot perform the statistical analysis. The score of these tests is increased. We have to carry out more experimental use but this result suggested that our learning environment is effective for

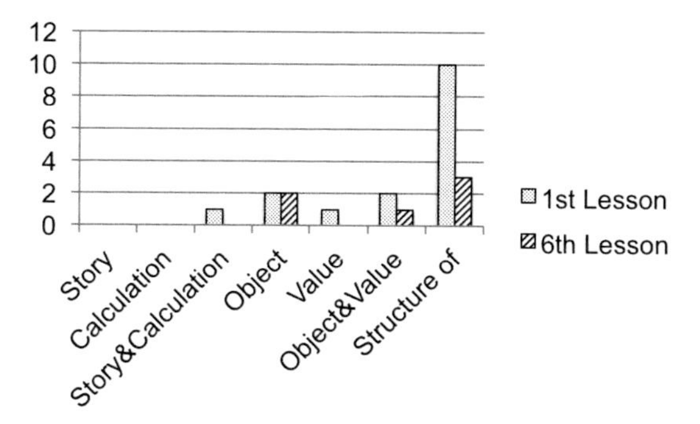

Fig. 9 Changing the rate of each error (5th grade student)

Table 1 Average score of pretest and posttest

	Pretest	Posttest
Usual problem solving	9.0	12.0
Extraneous problem solving	7.5	11.5

subjects who has LD to learn by posing the arithmetic word problems that can be solved by one-step addition or subtraction.

6 Summary and Future Works

In this paper, we have described about the design and experimental use of learning environment by posing problem for Learning disability. We have developed the learning environment for problem posing toward students in regular class. We developed this learning environment by redesigning the problem posing based on information structure of arithmetic word problem. And, a learner who has learning disability cannot pose the problem so it is impossible for them to learn by problem posing. However, we assumed that our learning environment except the cause of this difficulty. We have investigated a characteristic of learning disability and carried out an experimental use. As a result, two subjects are able to pose the problem and improve their problem solving performance.

As the future works, we have to perform more experimental use. We also have to analyze the characteristic of language delay students and improve this learning environment. Moreover, development other software based on same method of design is important works.

Acknowledgments I am deeply grateful to the teacher and the students in the elementary school for performing the experimental use and giving insightful comments.

References

1. Dougiamas, M., Taylor, P.: Moodle: using learning communities to create an open source course management system. In: Proceedings of the World Conference on Educational Multimedia, Hypermedia and Telecommunications, pp. 171–178 (2003)
2. McAuley, A., Stewart, B., Siemens, G., Cormier, D.: The MOOC model for digital practice (2010). http://www.davecormier.com/edblog/wp-content/uploads/MOOC_Final.pdf. Accessed 09 Sept 2015
3. Chang, K.E., Wu, L.J., Weng, S.E., Sung, Y.T.: Embedding game-based problem-solving phase into problem-posing system for mathematics learning. Comput. Educ. **58**(2), 775–786 (2012)
4. Fu-Yun, Yu.: Scaffolding student-generated questions: design and development of a customizable online learning system. Comput. Hum. Behav. **25**(5), 1129–1138 (2009)
5. Yamamoto, S., Kanbe, T., Yoshida, Y., Maeda, K., and Hirashima, T.: A case study of learning by problem-posing in introductory phase of arithmetic word problems. In: Proceedings of ICCE2012, Main Conference E-Book, pp. 25–32 (2012)
6. Polya, G.: How to solve it: A new aspect of mathematical method. Princeton University Press, Princeton (1957)
7. Silver, E.A., CAI, J.: An analysis of arithmetic problem posing by middle school students. J. Res. Math. Educ. **27**(5), 521–539 (1996)
8. Bender, W.N.: Learning Disabilities: Characteristics, Identification, and Teaching Strategies, 6th ed. Boston Pearson, Allyn and Bacon (2007)
9. Jitendra, A.K., Hoff, K.: The effects of schema-based instruction on the mathematical word-problem-solving performance of students with learning disabilities. J. Learn. Disabil. **29**(4), 422–431 (1996)
10. Hirashima, T., Hayashi, Y., Yamamoto, S.: Triplet structure model of arithmetical word problems for learning by problem-posing. In: Proceedings of HCII2014 (LNCS 8522), pp. 42–50 (2014)
11. Muth, K.D.: Extraneous information and extra steps in arithmetic word problems. Contemp. Educ. Psychol. **17**, 278–285 (1992)

An Approach to Estimating Decision Complexity for Better Understanding Playing Patterns of Masters

Akira Takeuchi, Masashi Unoki and Hiroyuki Iida

Abstract This paper proposes a method for estimating decision complexity of positions by using correlation coefficient between two evaluation values of the root node and leaf level of a game tree. Search is performed to determine the evaluation value at the root node with focus on the ratio between positive and negative values of position scoring at leaf nodes. Moreover, Kalman filter is employed to minimize the variance of scoring errors. Several games, played by masters and computers in the domain of shogi, are analyzed while applying the proposed method. The results show that the proposed idea is a promising way to better understand playing patterns of masters, for example, identifying the right moment for possibility of changing strategies such as speculative play and early resignation.

Keywords Decision complexity · Correlation coefficient · Kalman filter · Playing patterns of masters · Shogi

1 Introduction

A characterization of games with respect to their position in the game space has been discussed. It was defined as a two-dimensional space with the state-space (there called search-space) complexity as one dimension and the decision complexity as the second dimension [1]. The notion of decision complexity was rather vague; later it was replaced by the notion of game-tree complexity [2].

A. Takeuchi (✉) · M. Unoki · H. Iida
School of Information Science, Japan Advance Institute of Science and Technology, 1-1 Asahidai Nomi, Ishikawa 923-1292, Japan
e-mail: akir.takeuch@jaist.ac.jp

M. Unoki
e-mail: unoki@jaist.ac.jp

H. Iida
e-mail: iida@jaist.ac.jp

R. Lee (ed.), *Applied Computing & Information Technology*, Studies in Computational Intelligence 619,
DOI 10.1007/978-3-319-26396-0_9

The notion of decision complexity is important to understand the depth of master's thinking in the domain of game playing. To our best knowledge, no other research has been done to establish the notion of decision complexity in minimax tree framework.

Difficulty of a position is closely related to decision complexity. It has been considered in various contexts. For example, Jansen [3] studied problematic positions and speculative play. He noted that somewhat counter-intuitively, whether a move deserves an exclamation make is not so much a characteristic of the move (and the resulting position) itself, but rather from the initial position from which the move was made. It seems logical therefore to define problematicity as a characteristic of a position rather than a move. This leads to a strongly related concept, the difficulty of a position.

Proof number (PN) [4] is related to the difficulty, which relates to the minimum number of unsolved nodes that need to be solved. So, maximum PN shows the complexity to solve these unsolved nodes. On the other hand, disproof number (DN) is related to the minimum number of unsolved nodes that need to be disproved. Therefore, maximum DN shows the complexity to disprove. In other word, both the maximum PN and maximum DN are an effective measures of difficulty to solve nodes as soon as possible for the AND/OR tree framework [5].

A well-known search indicator studied by several researchers in the late-80s is the Conspiracy-Number Search (CNS). Introduced by McAllester [6, 7], CNS is a best first search algorithm for minimax tree frame-work, which determines the cardinality of the smallest set of leaf nodes which have to "conspire" to change their values in order to change the minimax value of the root. The maximum Conspiracy-Number (MaxCN) and minimum Conspiracy-Number (MinCN) show a correlation to maximum PN and maximum DN in the minimax tree framework [8].

In the CNS context, MaxCN and MinCN identify the critical position in the game's progress for an expected outcome. MaxCN indicates the unlikeliness of root value in achieving a value. Therefore, high value of MaxCN implies the high likeliness of winning or losing. MinCN indicates that the likeliness of root value to achieve a value is limited to the value of MinCN. So, high value of MinCN implies the possibility of target change of the possible estimated outcome. Considering the relation to maximum PN and maximum DN, we find that high values of MaxCN and MinCN are effective measures of difficulty of a particular value to be likely.

Khalid et al. [9] suggested the possibility to identify critical positions through MaxCN and MinCN. Why do we use CNS instead of the positional scoring (evaluation function) alone during the game's progression? The positional scoring acts as a guiding function for an individual player to determine his/her state in the game's progress. Although it might be suitable for identifying critical positions, it still lacks the probabilistic element in abstracting the player's decision to determine the next game state. Basically, positional scoring shows how the game progresses (i.e. which player is leading the game) but the CN-values potentially show its probable changes. However, we understand through our experience that it is practically hard for master-level (deeper search) computer shogi to determine conspiracy numbers.

The following observation was also made through our experience. When searching a position with some advantage for one's side, it is likely that evaluation values at leaf nodes of game tree are also often positive, i.e., good for that player. Conversely, when disadvantageous positions are dominant at leaf nodes, it is likely that the root score may fail low. Inspired by the observation, we propose an idea to use the correlation coefficient, as an estimation of decision complexity, between the root node and leaf level. The value at leaf level is calculated by the ratio between positive and negative values of position scoring. Then, Kalman filter is incorporated to minimize the variance of scoring errors. Analyzing game scores played by masters and computers while applying the propose method, results in an interesting outcome to better understand the master's play such as recognition of resignation timing at a position where it looks still even.

The structure of the paper is as follows. Section 2 describes our proposed method for estimating decision complexity of positions. Section 3 shows some analysis examples using game scores played by masters and computers. Section 4 gives some discussions and the concluding remarks are given in Sect. 5.

2 An Approach to Estimating Decision Complexity

This section presents the procedures for estimating decision complexity derived from game scores played by masters. The procedures consist of measurement from game scores, time-series analysis using Kalman filter, and estimation by using correlation coefficient.

2.1 Measurement from Game Scores

A simple way to measure the ratio between positive and negative values of position scoring at the leaf nodes is only to record that the evaluation value is positive or negative. This can be easily implemented in most chess-like computers using alpha-beta algorithm.

Each position which appears in the given game scores used for analysis is searched sequentially. Search for each position is aborted when the number of nodes reaches 2^{30}. This corresponds to the number of nodes that can be searched in average thinking time under the tournament condition where one player takes 4–5 hours per game.

2.2 Time-Series Analysis Using Kalman Filter

By the above-mentioned measurement method, the set of evaluation value at root node and the ratio between positive and negative values of position scoring at the leaf nodes are obtained for each position in the given game scores. The measured values of these two evaluation factors may include errors due to the evaluation function and search aborting. Therefore, we estimate the evaluation value, its velocity, and its acceleration by minimizing the variance of the errors using Kalman filter.

The ratio $r(x)$, such as win probability corresponding to the evaluation value x, can be represented as follows

$$r(x) = 1/(1 + \exp(-x/T))$$

where T is constant value depending on the evaluation function and is adjusted to 256 in our computer program used in this study. By this equation, the two measured values can be converted into comparable factors each other. Here, the ratio between positive and negative values r is reverse converted to the state variable $x(r)$ corresponding to the evaluation value by the following equation.

$$x(r) = -T\ln(1/r - 1)$$

In the following, we call this converted value as the evaluation value at leaf level. The reason why we use the reverse conversion is that there is no indication of process noise and observation noise, which are required in the application of Kalman filter, in the ratio. On the other hand, the observation noise for the state variables based on the evaluation value can be estimated from the margin for pruning such as ProbCut [10].

In the following application of Kalman filter, time corresponds to the ply and each position for ply Δt is searched.

State variable vector X_k consists of evaluation value, its velocity, and its acceleration. State equation is expressed using the state of previous position X_{k-1} as follows.

$$X_k = FX_{k-1} + Gw_k$$

where

$$X_k = \begin{bmatrix} x_k & \dot{x}_k & \ddot{x}_k \end{bmatrix}^T$$

$$F = \begin{bmatrix} 1 & \Delta t & \frac{\Delta t^2}{2} \\ 0 & 1 & \Delta t \\ 0 & 0 & 1 \end{bmatrix}$$

$$G = \left[\frac{\Delta t^2}{2} \quad \Delta t \quad 1\right]^T$$

$$w_k \sim N(0, Q_k)$$

The acceleration applied to the state variables contains noise w_k, which is assumed to follow a normal distribution of variance Q_k.

When y_k is an observed value or its converted value, the observation equation is expressed as follows.

$$y_k = HX_k + v_k$$

where

$$H = [1 \quad 0 \quad 0]$$

$$v_k \sim N(0, R_k)$$

Here, v_k is observation noise which is assumed to follow a normal distribution of variance R_k. The variance Q_k and R_k are set from the margin which depends on the position.

Prediction procedures in the analysis method using Kalman filter are expressed as follows:

$$X_{k|k-1} = FX_{k-1|k-1}$$

$$P_{k|k-1} = FP_{k-1|k-1}F^T + GQ_kG^T$$

where $X_{k|k-1}$ is the state estimate of position k at the time $k - 1$, and $P_{k|k-1}$ is the covariance matrix of the error.

Furthermore, the update procedures are expressed as follows:

$$e_k = y_k - HX_{k|k-1}$$

$$S_k = R_k + HP_{k|k-1}H^T$$

$$K_k = P_{k|k-1}H^T S_k^{-1}$$

$$X_{k|k} = X_{k|k-1} + K_k e_k$$

$$P_{k|k} = (I - K_k H)P_{k|k-1}$$

where e_k is observation residual error, S_k is covariance of observation residual error, and K_k is Kalman gain.

By iterating prediction and update in accordance with the progress of the game, the analysis is performed. Although there is a correlation between the evaluation value at root node and leaf level, these state variables are analyzed independently because it is difficult to formulate the relationship.

2.3 Estimating Decision Complexity Using Correlation Coefficient

As one method of estimating decision complexity, we perform the correlation analysis between the two evaluation factors. The formula for correlation coefficient ρ is as follows:

$$\rho = \frac{\sum_{k=1}^{n}(x_{Rk} - \overline{x_R})(x_{Lk} - \overline{x_L})}{\sqrt{\sum_{k=1}^{n}(x_{Rk} - \overline{x_R})^2}\sqrt{\sum_{k=1}^{n}(x_{Lk} - \overline{x_L})^2}}$$

where x_{Rk} and x_{Lk} are evaluation values at root node and leaf level, $\overline{x_R}$ and $\overline{x_L}$ are each of the mean values, and n is the number of sample positions.

In addition to the calculation of the correlation coefficient for all positions in the game, we analyze the changing process of the correlation coefficient by limiting to 16 consecutive positions in the following section.

3 Game Analysis by Decision Complexity

We analyze several game scores of title match in shogi while applying the proposed method. As described in the previous section, game-tree search by the thinking time under the title match condition were performed for the measurement from game scores. We then observe important aspects of master's play and confirm validity of the proposed method.

3.1 First Game of the 62nd Meijin Title Match (A)

We analyze the game played between the reigning champion Mr. Habu and challenger Mr. Moriuchi in the 62nd Meijin (most prestigious) title match in April 2004. This game has shown a puzzling early resignation which highly surprised the spectators.

Figure 1a shows the correlation coefficient between the evaluation values at root node and leaf level. After the correlation coefficient falls in the middle game, it continues to rise closer to 1 toward the end of the game or resignation. This tendency is more clearly reflected in the correlation coefficient of the estimated evaluation values with Kalman filter, which shows effectiveness of the proposed

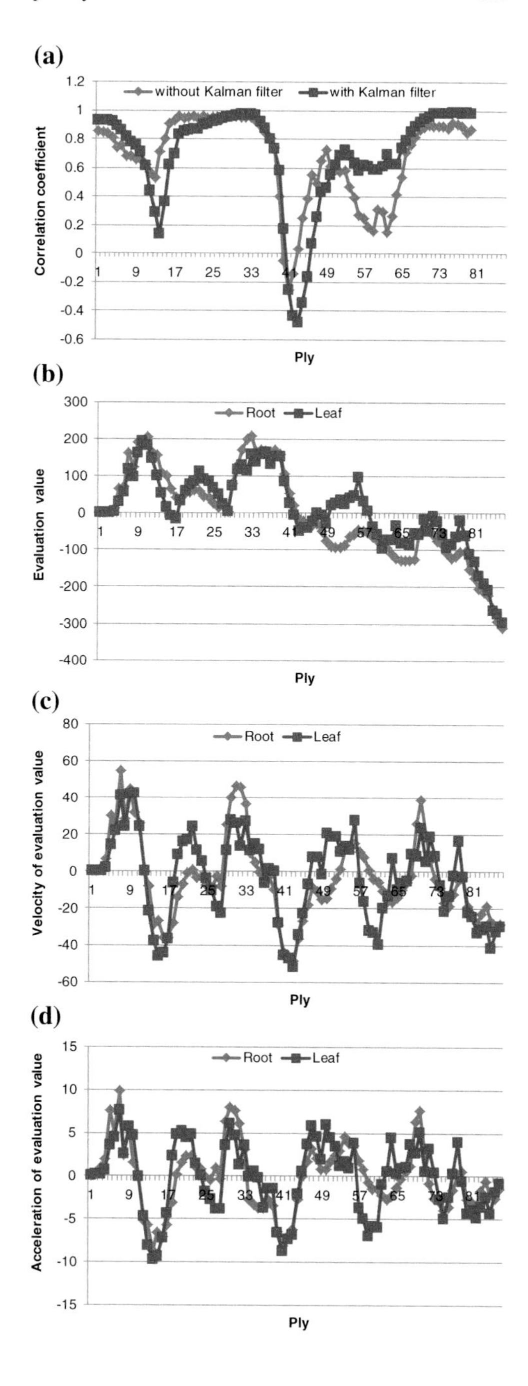

Fig. 1 Correlation coefficient and state variables against game progress: **a** correlation coefficient; **b** evaluation value; **c** velocity of evaluation value; **d** acceleration of evaluation value

method. Figure 1b–d show state variables, its velocity, and its acceleration, respectively. Each figure compares the evaluation value at root node and leaf level against game progress. Focusing on near resignation, velocity and acceleration values continue to take negative ones.

The analysis results may indicate that the player had to resign because the position became easy to understand. It should be noted that even strongest computer shogi cannot understand the reason for such early resignation without focusing on decision complexity or its estimation. The masters-like resignation can be implemented by the identification such that the correlation coefficient becomes higher than the players specific threshold.

3.2 Fourth Game of the 60th Ouza Title Match (B)

The second game to analyze was played by the reigning champion Mr. Watanabe and challenger Mr. Habu in the 60th Ouza title match in October 2012. This game was selected as first place of 2012 best match by professional shogi players, whereas the game ended in a draw.

The low correlation coefficients shown in Fig. 2 indicate that there are difficult positions in the opening, middle and end game. In particular, positions around the amazing move at position of 122nd ply can be determined to be very difficult, because there is no correlation of the estimated evaluation values between the root node and leaf level. Thus the analysis results are consistent with the assessment by professional players.

3.3 First Game of the 2nd Denou Title Match (C)

This game was played between GM Mr. Abe and computer program *Shueso* in the 2nd Denou title match in March 2013. In this game, the anti-computer strategy was successful and Mr. Abe won by letting off the attack of the computer.

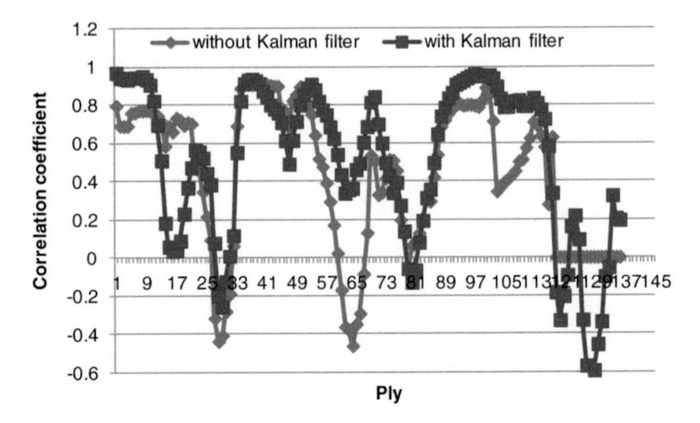

Fig. 2 Correlation coefficient against game progress

Figure 3b–d show that tendency of evaluation values at root node and leaf level is similar except for the part of the endgame. The velocity continues to take positive value after the position of 100th ply. Looking at correlation coefficient shown in Fig. 3a, values with Kalman filter are closer to 1 in the endgame.

The appearance of positions to be understood easily is the characteristics for the game played by computer. The computer using the resignation identification described above must be able to resign at earlier ply.

3.4 First Game of the 3rd Denou Title Match (D)

We continue to analyze the game played between GM Mr. Sugai and computer program *Shueso* in the 3rd Denou title match in March 2014. In this game, even though there is no clear bad move, computer was able to expand advantage little by little after the middle game.

Figure 4 shows that the middle game was very difficult because of lower correlation coefficient with Kalman filter. The analysis result is consistent with the assessment by masters. For example, *Shueso* played a surprising good move at the position of 50th ply, which was highly recognized by a GM in charge for giving annotation to the game.

3.5 Rematch After the 3rd Denou Title Match (E)

Finally, we analyze a rematch game played also between GM Mr. Sugai vs. computer program *Shueso* in July 2014 after the 3rd Denou title match. This game appears to have been well balanced until the endgame and became exciting as expected.

Figure 5b shows that the evaluation value at the root node and leaf level is contradictory during a long period in the endgame. This feature appears to the changes of the correlation coefficient as shown in Fig. 5a, and positions estimated to be more difficult also appear in the middle game. There were nearly an hour of thinking by the computer and more than 3 h of long thinking by GM at positions of 74th and 75th ply, respectively. The computer log shows that the evaluation value was varying and best move was replaced many times during the long thinking. Therefore, there are very difficult positions in the middle game. It supports the analysis result of the correlation coefficient with Kalman filter which becomes clearly lower in the middle game.

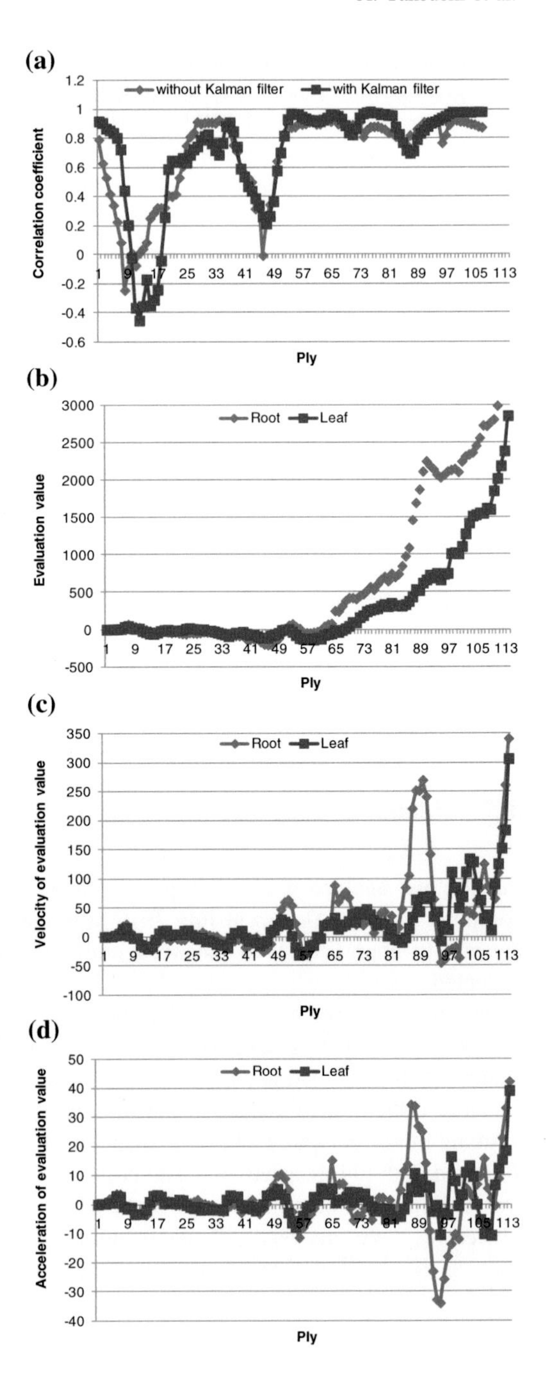

Fig. 3 Correlation coefficient and state variables against game progress: **a** correlation coefficient; **b** evaluation value; **c** velocity of evaluation value; **d** acceleration of evaluation value

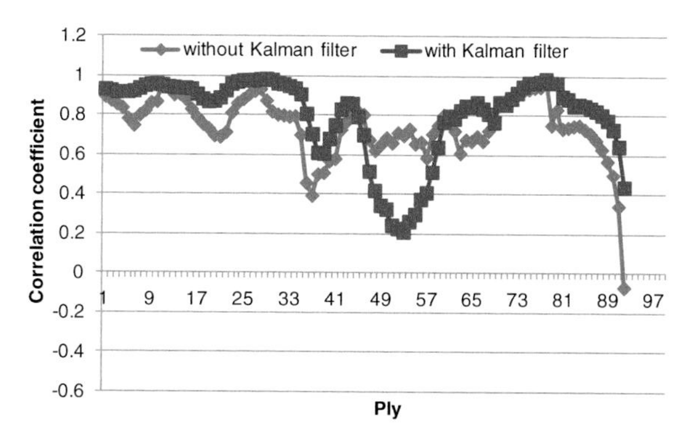

Fig. 4 Correlation coefficient against game progress

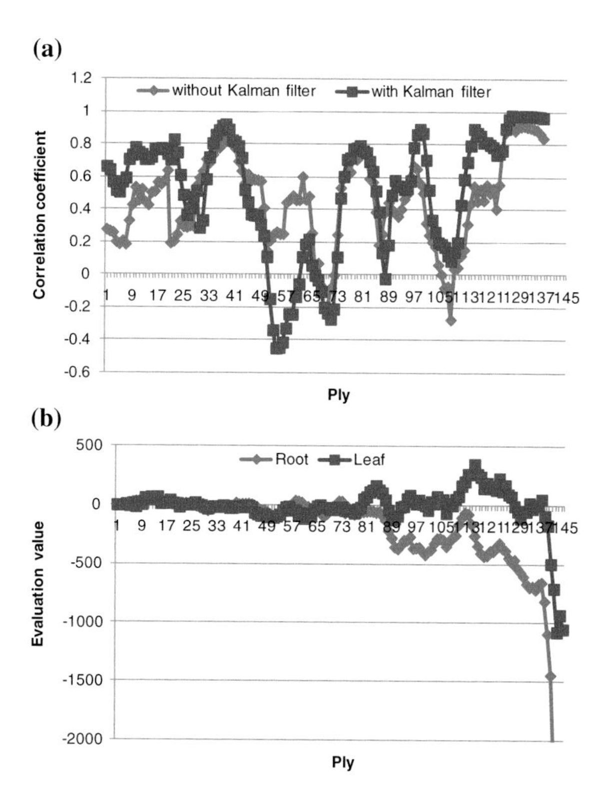

Fig. 5 Correlation coefficient and evaluation value against game progress: **a** correlation coefficient; **b** evaluation value

4 Discussion

Figure 6 shows the relation between the correlation coefficients and the number of searched nodes. The correlation coefficient calculated from the measured evaluation values shown in Fig. 6a is lower than that calculated from the estimated evaluation values with Kalman filter shown in Fig. 6b. In game scores (B) and (E) where difficult positions are dominant, the differences are clear and the correlation

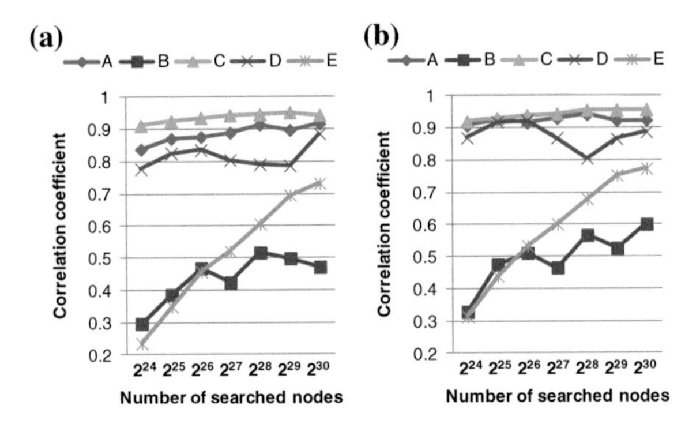

Fig. 6 Correlation coefficient for the number of searched nodes: **a** correlation coefficient of the measured evaluation values without Kalman filter; **b** correlation coefficient of the estimated evaluation values with Kalman filter

Table 1 Correlation coefficient and root-mean-square speed throughout the game scores

Game score	Correlation coefficient (without Kalman filter)	Root-mean-square speed
(A) First game of 62nd Meijin title match	0.923 (0.916)	23.2
(B) Fourth game of 60th Ouza title match	0.600 (0.470)	27.1
(C) First game of 2nd Dennou title match	0.956 (0.943)	52.4
(D) First game of 3rd Dennou title match	0.886 (0.885)	31.6
(E) Rematch after 3rd Dennou title match	0.773 (0.732)	44.6

coefficient becomes higher as the number of searched nodes increase. This is a natural relationship that the degree of difficulty is lowered in accordance with that the thinking ability becomes higher. In game scores (B) and (E), the estimated evaluation values for the number of searched nodes 2^{28} corresponds to the measured evaluation values for the number of search nodes 2^{30}. Therefore, the proposed method can estimate the evaluation values with higher accuracy using Kalman filter as well as increasing the searched nodes.

Table 1 shows the correlation coefficient and root-mean-square speed throughout the game scores. As described above, lower correlation coefficient indicates that difficult positions are dominant. High correlation coefficient is shown in the game (A) that became disappointing contents as Meijin title match by early resignation, and in the game (C) in which the result may be determined in the middle game. On the other hand, small value of root-mean-square speed indicates the high degree of perfection, and its large value is considered to indicate that the game was exciting.

As this hypothesis, the root-mean-square speed has a small value in the game (A) and (D) that there was no clear bad move, and in the game (B) in which the result was draw. Therefore, the assessment similar to experts can be made by the proposed estimation of position difficulty or decision complexity.

Statistical considerations using more samples of game scores are required, and various factors are necessary for artistic assessment of games. However, correlation analysis of the proposed evaluation factors can be a powerful method for artistic assessment of games considered.

5 Conclusion

With the aim to obtain more insightful information than positional scoring in minimax tree framework, we proposed a method of estimating the difficulty of position or decision complexity. Several games, played by masters and computers in shogi, were analyzed by applying the proposed method. As the results, it was found that the ratio between positive and negative values of position scoring at leaf nodes can be a practicable factor for decision complexity, and effective information can also be obtained from the velocity and acceleration of this factor. Furthermore, the play of masters and the difficult or interesting positions could be identified by the correlation coefficient of the estimated evaluation values with Kalman filter more clearly.

Since the proposed evaluation factors can be measured with cheap computational cost, it is easy to analyze online during the progress of the game. The proposed method has the potential to be applied in various ways. For example, a speculative play might be carried out while selecting a move to induce decision complexity. Moreover, estimating decision complexity makes it possible to consider strategic change as soon as possible, for example, from winning to drawing.

Acknowledgements The authors wish to thank the anonymous referees for their constructive comments that helped to improve the article considerably. This research is funded by a grant from the Japan Society for the Promotion of Science in the framework of the Grant-in-Aid for Challenging Exploratory Research (grant number 26540189).

References

1. van den Herik, H.J., Uiterwijk, J.W.H.M., Rijswijck, J.V.: Games solved: now and in the future. Artif. Intell. **134**(1–2), 277–311 (2002)
2. Allis, L.V.: Searching for solutions in games and artificial intelligence. Ph.D. thesis, University of Limburg (1994)
3. Jansen, P.: Problematic Positions and Speculative Play, Computers, Chess and Cognition, pp. 169–182 (1990)
4. Allis, L.V., van der Meulen, M., van den Herik, H.J.: Proof-number search. Artif In-tell **66**(1), 91–124 (1994)

5. Ishitobi, T., Cincotti, A., Iida, H.: Shape keeping technique and its application to checkmate problem composition. In: Ninth Artificial Intelligence and Interactive Digital Entertainment Conference. Workshop (WS-13–19), 7–10 (2013)
6. McAllester, D.A.: A new procedure for growing min-max trees. Technical report, Artificial Intelligence Laboratory, MIT, Cambridge, MA, USA (1985)
7. McAllester, D.A.: Conspiracy numbers for min-max search. Artif. Intell. **35**(3), 287–310 (1988)
8. Hashimoto, J.: A study on game-independent heuristics in game-tree search. Ph.D. thesis, School of Information Science, Japan Advanced Institute of Science and Technology (2011)
9. Khalid, M.N.A., Yusof, U.K., Iida, H., Ishitobi, T.: Critical position identification in games and its application to speculative play. ICAART **2015**(2), 38–45 (2015)
10. Buro, M.: ProbCut: an effective selective extension of the alpha-beta algorithm. ICCA J. **18**(2), 71–76 (1995)

A Routing Algorithm for Distributed Key-Value Store Based on Order Preserving Linear Hashing and Skip Graph

Ken Higuchi, Makoto Yoshida, Naoyuki Miyamoto and Tatsuo Tsuji

Abstract In this paper, a routing algorithm for the distributed key-value store based on order preserving linear hashing and Skip Graph is proposed. In this system, data are divided by linear hashing and Skip Graph is used for overlay network. The routing table of this system is very uniform. Then, short detours can exist in the route of message forwarding. The proposed algorithm detects such detours and changes the route. By using the proposed algorithm, reduction of the number of hops are expected. From experimental results, it proved that the proposed algorithm is effective.

Keywords Distributed key-value store · Skip graph · Linear hashing

1 Introduction

In recent years, cloud computing become popular tool in business and personal areas, and using data stored in a faraway place is not special now. On the other hand, the amount of data that is held by companies and individuals has increased greatly. Therefore, it is natural that the data is distributed and managed in many individual computers (nodes). These distributed systems are researched in many years. One of the solutions is distributed database systems. But the distributed database system based on the conventional database system such as a relational database (RDB) has some problems. In order to ensure ACID properties of transactions, various restrictions are added to the database system, then, the performance of it is degraded. In the distributed system, this deterioration is remarkable. Then many users want to use simply and comfortably without stringent. In order to meet

K. Higuchi (✉) · M. Yoshida · N. Miyamoto · T. Tsuji
Graduate School of Engineering, University of Fukui, Fukui, Japan
e-mail: higuchi@u-fukui.ac.jp

T. Tsuji
e-mail: tsuji@u-fukui.ac.jp

R. Lee (ed.), *Applied Computing & Information Technology*,
Studies in Computational Intelligence 619,
DOI 10.1007/978-3-319-26396-0_10

these demands, distributed key-value stores (Distributed KVS) systems have attracted attention, and are used for various services. Key-value stores (KVS) systems are scale-out easily store data model by limiting the format of data in the key-value pair and limiting possible operations and functions. The distributed KVS is adaptable to the large-scale data by storing the data in a distributed manner to a plurality of nodes on the network.

Many existing distributed KVS such as Amazon Dynamo [2] and Apache Cassandra [8] use a consistent-hashing [6] for the key to divide and store the data to the nodes. Also, in order to reach the node to be processed in 1-hop forwarding, the gateway nodes and the client nodes have to hold the entire routing information to each node. Thus, it is necessary to share the routing information in a plurality of nodes and it requires expensive process at the time of change of the routing information. It cannot be said that they are efficient to insert or delete a node.

On the other hand, by using Distributed Hash Table (DHT) and Skip Graph [1], it is also conceivable to build the distributed KVS. Chord [12] and Kademlia [10] are popular DHT. In these methods, a query is transmitted to the node in multi-hop forwarding. Moreover, all nodes have the same function, and the node decides a node to forward in order to send the query to the target node of the query. That is, each node is not required to have the entire routing information for all nodes, and has only a portion relating to itself. The entire routing information is held on a whole node set. In this method, when inserting or deleting a node, it is not necessary to update the routing information in all nodes and it is only necessary to the corresponding routing information in some nodes. Then, the routing information is updated efficiently. However, on the DHT that determines the data storage location by using the hash function, although excellent in load balancing, the order of the original key in the nodes is often not preserved. Thus, for a range search, it is necessary to transfer the query to all nodes, and very inefficient.

In contrast, Skip Graph [1] is an overlay network that is possible range search. Furthermore, there are some extensions of Skip Graph. Multi-key Skip Graph [7] is extended Skip Graph for efficient range retrieval. On original Skip Graph, a node can handled only one key, but on Multi-key Skip Graph, a node can handled multiple keys by using virtual nodes. By using this Skip Graph, the distributed KVS can be implemented. But the routing table becomes to be large because the routing table is necessary for each key. Then, the cost of node insertion and node deletion is very high. Range-Key Skip Graph [4] is another extension of Skip Graph. It simplifies the routing table of Multi-Key Skip Graph. In addition, a large-scale distributed KVS using Range-key Skip Graph has also been proposed [5]. However, it is not performed efficiency when a node is inserted or deleted.

Distributed KVS based on Order Preserving Linear Hashing and Skip Graph [13] is another implementation of the Distributed KVS. In this system, the data is divided by Order Preserving Linear Hashing(OPLH) and its overlay network is Skip Graph. This system can easily insert and delete the node and the number of hops of query forwarding is almost the same as that of [5].

In this paper, we propose a new routing algorithm for the distributed KVS based on OPLH and Skip Graph. It uses detours of the route, and the number of hops of query forwarding is reduced. By experiments, efficiency of our new algorithm is proved.

2 Distributed KVS Based on Order Preserving Linear Hashing and Skip Graph

2.1 Order Preserving Linear Hashing

Linear hashing [9] is a kind of dynamic hashing. It consists of a hash function, a bucket array (hash table), data buckets, and meta-information. Meta-information includes the hash level, the number of data buckets, the number of records, and a fixed threshold. The data bucket is labeled by the different non-negative integer that is less than the number of data buckets. These labels correspond to hash values, and each data bucket stored only the set of data whose hash values are its label. The bucket array keeps the addresses of data buckets in order of hash values. In general, the hash function is modulo by 2^i for hash level i; i.e. the hash value is the i bits suffix of the bit pattern of the data.

In the process of data retrieval, the hash value of the target data is calculated by the hash function (whose hash level is i), and the address of the data bucket storing the target data is searched from the bucket array. If the calculated hash value is not less than the number of data buckets, the data bucket corresponding to this hash value doesn't exist. In this case, the hash value is recalculated by the hash function whose hash level is $i - 1$. Then the data bucket corresponding to the recalculated hash value is searched for the retrieval. When the ratio of the number of record to the number of data buckets is greater than the threshold, the bucket array is expanded to have a reference to a new data bucket. The hash value of the new data bucket is the number of data bucket minus 1 and is greater than the other hash values. But if the number of data buckets is 2^i, the corresponding hash value of the new data bucket is already used. When the number of data buckets becomes to be greater than 2^i for hash level i, the hash level is increased by 1. Then, on the linear hashing whose hash level is i, the hash level corresponding to the data bucket is i or $i - 1$.

Order Preserving Linear Hashing [3, 11] (OPLH) is a linear hashing which uses the special hash function. It employs a combination of the division function and the bit reversal function. This function outputs the reverse ordered bit string of the result of the division function. For examples in hash level is 3, the hash value of 010101 is 010 and the hash value of 011000 is 110.

On OPLH, the expansion method and the reduction method for the bucket array are the same as that of the linear hashing using modulo function. When hash level is incremented, the old hash values are added “0” as the prefix and the hash value of the new element of the bucket array is greater than the other hash values. Then OPLH inherits the advantage of the traditional linear hashing with good response time for the range query.

2.2 Skip Graph

Skip Graph [1] is a distributed data structure, based on skip lists. It is a kind of structured overlay network, and it supports range retrievals. It is suited to peer-to-peer networks. Unlike DHT, Skip Graph doesn't use a hash function to decide to data partitioning. On Skip Graph, the data is clustered by the range of key similar to the sequence set of B^{+}tree and the order of keys is kept. Then, range retrievals are processed effectively.

Figure 1 shows an example of Skip Graph. In Fig. 1, the group of rectangles that have same number is a node and the number of the group of rectangles is the key of data that are stored by the node. Each rectangle represents an entry in the routing table and a link between a pair of rectangles expresses that it is a valid entry. Each entry of the routing table belongs to one level of Skip Graph. Each node has the randomized binary number which called a *membership vector*. Membership vectors are used for the routing table of Skip Graph. In level 0 layer, the nodes are sorted by keys and only the adjacent nodes are connected. In other level i, the rectangles are partitioned by the first i bit of the membership vector, only the adjacent nodes in such partition are connected. Each valid entry of the routing table includes the node address and the key. Here, for the number of nodes N, the maximum level is $\lceil \log_2 N \rceil$.

On Skip Graph, the route from the node A (whose key is a) to another node B (whose key is b) is decided as followings. Here, $a < b$ is assumed. Firstly the highest key that is not more than b and its node are searched from the routing table of A. Let such key be c and let C be the node whose key is c. Next, node A forwards the message to node C and C forwards this message to the next node similarly. By repeating this forwarding, the message is reached to the node B. Here, this routing algorithm used routing tables and key (b) can decide the route for exact retrieval and range retrieval.

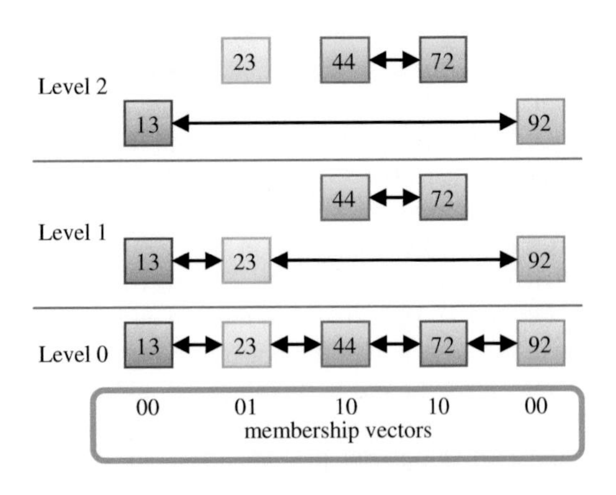

Fig. 1 An example of Skip Graph

2.3 Distributed KVS

Distributed KVS based on OPLH and Skip Graph was proposed [13]. In this system, the data is partitioned by OPLH and Skip Graph is used for the overlay network. In other words, the buckets of OPLH correspond to the nodes. But the hash values are not only used for the data partitioning but also the membership vectors and the labels of nodes. Then this system has the following features.

- The set of the node labels is that of consecutive non-negative integers started from 0.
- In level 0 layer, the nodes are not sorted by their stored keys.
- In level i, the node labeled a is linked to the nodes whose labels are $a \pm 2^i$ if exist.

Figure 2 shows an example of Skip Graph based of OPLH. In Fig. 2, the binary number is the node label and also the hash value. Its length is the hash level. Each node has the label with the hash level.

2.4 Routing Algorithm

In this system, the routing is similar to the binary search. In one routing, the message is not forwarded twice or more using same level of Skip Graph. Then, maximum number of hops of the message forwarding is not more than the maximum level of Skip Graph +1 ($\lceil \log_2 N \rceil$ for the number of nodes N). But for retrieval, the calculating the hash value is difficult on this system. In OPLH, the hash value is calculated by using the hash level and the number of buckets. Then, if some node wants to calculate the correct hash value, the node has to use the hash level and the number of nodes. But these are global information and the node can only use to the local hash level of the node in many cases. Here, the local hash level means the hash level for the node such as the bucket of OPLH. Then, the following cases occur. Here the node that is received query by client is called the *start node* and the node that stores result data for the query is called the *target node*.

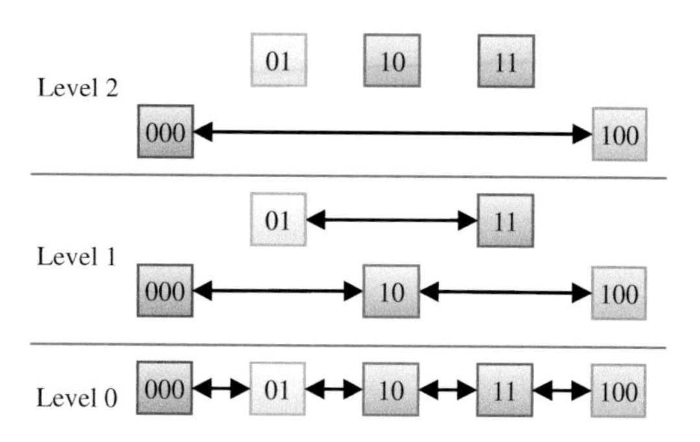

Fig. 2 Skip Graph based on OPLH

- The node of the hash value that is calculated in the start node does not exist because start node cannot revise using the number of nodes.
- The node of the hash value that is calculated in the start node exists but it is not the target node because the local hash level of the start node is not equal to the global hash level.

In order to solve these cases, this system uses the following revision using the maximum label of node in the routing table in the start node.

- If the calculated hash value is more than the maximum label in the routing table of the start node, the hash value is recalculate using the hash function whose hash level is local hash level minus 1.

By this revision, the first case of above cases doesn't occur, but second case still occurs. To solve the second case, after reaching to the node whose label is calculated in the start node, this node checks that the key of the query is included in the range of keys that are stored in this node. If not, this query is forwarded to the correct target node after the recalculation in this node. Here, this node is linked to the correct target node in the routing table of this node, and then the number of hops of this forwarding is only 1.

Figure 3 shows the route from the node 000 to the node 11. Here Let the node 000 be the start node and the key of the query be 111. Firstly, the hash value is calculated in the node 000. Because the hash level of the node 000 is 3, the hash value of 111 is 111. But the maximum label in the routing table of the node 000 is 100 and the hash value is revised. The revised hash value is 11 and the query is forwarded through the route 000 $\rightarrow$ 10 $\rightarrow$ 11. In case that the node 111 exists, the query should be forwarded to the node 111. But the node 111 is linked by the node 11(to be exact, 011) in the routing table of the node 11. Then, by 1-hop forwarding, the query reaches to the target node.

2.5 Node Insertion and Deletion

The method of the node insertion/deletion is similar to the expansion/reduction method for the bucket array of OPLH. But there is the problem in node deletion.

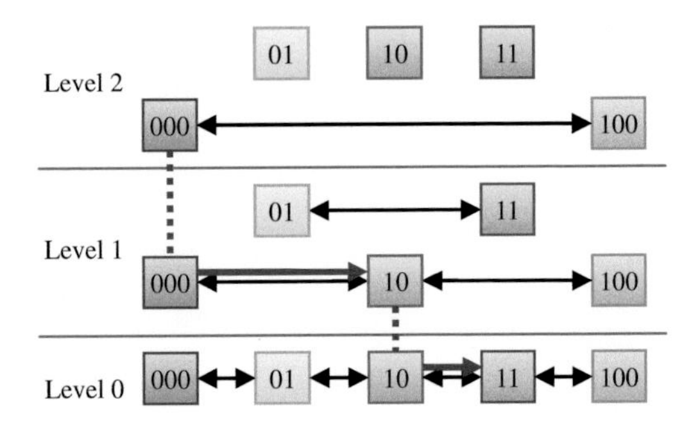

Fig. 3 The route from 000 to 11

In OPLH, the target data bucket of the reduction is the data bucket whose label is greatest. But in the peer-to-peer system, the target node of the deletion is not only the node whose label is greatest. Then, in this system, the node whose label is greatest is deleted firstly (including the data migration and the modification of the routing table), whole data of the target node of the deletion are moved to that node (deleted from the system already) and routing tables of the corresponding nodes are modified. By this method, any node can be deleted from the system.

3 Routing Algorithm Using the Detour

The routing algorithm described in Sect. 2 is based on Skip Graph. But the routing table of [13] is very uniform. Then the route from some node to another node using Skip Graph can be calculated. Furthermore, if the node knows the global hash level, that node can calculate the shortest route. By using these features, the number of hops of message forwarding can be reduced. In this section, we propose a new routing algorithm suitable for the distribute KVS based of OPLH and Skip Graph. In our algorithm, the short detour of the route decided by Skip Graph is detected and the route is changed to the detour. Then, the number of hops of message forwarding is reduced in many cases.

3.1 *Concept of the Detour*

The route calculated by Skip Graph has the feature that the directions of the direct routes in the route are the same to each other; i.e. they are only increasing directions or only decreasing directions for the node labels, and they are not the mix of increasing directions and decreasing directions. If a mixed route of increasing directions and decreasing directions can be used then the number of hop of message forwarding may be reduced. Figure 4 shows an example of detours. In Fig. 4, the

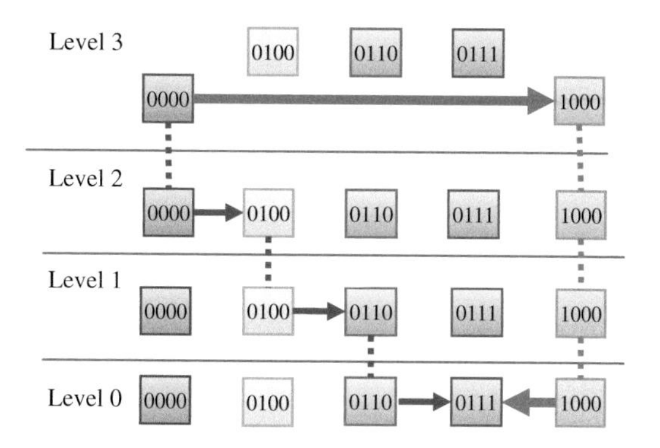

Fig. 4 The detour from 0000 to 0111 (the nodes 0001, 0010, 0011, and 0101 are omitted)

red route (3 hops) is calculated by Skip Graph and the blue route is the short detour (2 hops). By using this detour, the number of hops from the start node to the target node can be reduced.

3.2 Detecting the Detour

Firstly, let $a_i \to a_j$ be the direct route from the node labeled a_i to the node labeled a_j, and the concatenation of direct routes is presented as $a_{n+1} \to a_n \to \cdots \to a_1 \to a_0$. This is route from the node labeled a_{n+1} to the node labeled a_0.

Let $(a_{n+1}, a_n, \ldots, a_1, a_0)$ be the route $a_{n+1} \to a_n \to \cdots \to a_1 \to a_0$ calculated by Skip Graph. Here, $a_{i+1} \to a_i$ means that the node labeled a_{i+1} has the valid entry of the routing table to a_i at level i of Skip Graph and the forwarding is occurred by using this entry. Here, $n+1$ is the local hash level of the node labeled a_{n+1}. In some routes calculated by Skip Graph, it holds that $a_{i+1} = a_i$. In this case, the node labeled a_{i+1} doesn't forward to another node using level i entry of the routing table. Let $(b_n, b_{n-1}, \ldots, b_1, b_0)$ be *a direction vector* of the route $(a_{n+1}, a_n, \ldots, a_1, a_0)$ where for any $0 \le i \le n, b_i \in \{1, 0, -1\}$. Here if $a_{i+1} \to a_i$ is increasing direction then $b_i = 1$, if $a_{i+1} \to a_i$ is decreasing direction then $b_i = -1$, and if $a_{i+1} = a_i$ then $b_i = 0$. Then

$$a_{n+1} + \sum_{i=0}^{n} b_i \times 2^i = a_0$$

because of the features of the distributed KVS based on OPLH and Skip Graph in Sect. 2.3. From above equation, even if the applying order of levels in the routing table, direction vector is not changed and the query can be forwarded from the start node to the target node (of course, the route is changed.)

The detour such as Sect. 3.1 is detected by using the direction vector. If the direction vector has the followings sub-part, the short detour exists. Here $b_{-1} = 0$ formally.

- There exist i and j $(i > j+2, j \ge -1)$ such that $b_i = b_j = 0$ and for any $k(i > k > j+1)$, it holds that $b_k = b_{k-1} \in \{1, -1\}$.

By changing this sub-part as following respectively, the direction vector becomes the detour.

- $b_i = b_{i-1}$, $b_{j+1} = -b_{i-1}$ and for each $k(i > k > j+1)$, $b_k = 0$.

But the above detour is not always correct because such detour forwards via the node that does not exist. For example, consider the detour from the node labeled $0111_{(2)}$ to the node labeled $0000_{(2)}$. The direction vector of this route is $(0, -1, -1, -1)$. By using above method, the direction vector of the detour is $(-1, 0, 0, 1)$ and the route of detour is $7_{(10)} \to -1_{(10)} \to -1_{(10)} \to -1_{(10)} \to 0$. But the node labeled -1 does

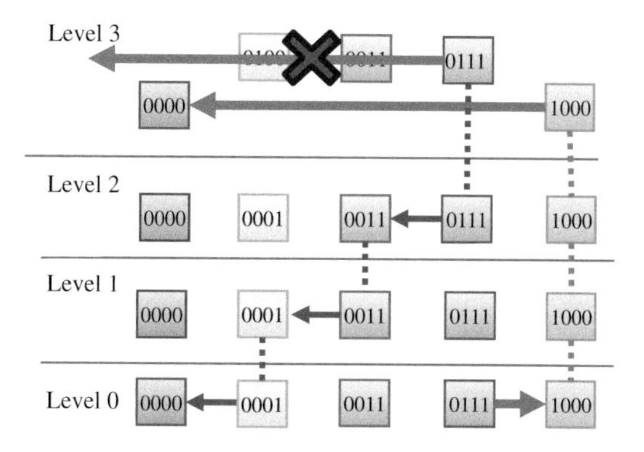

Fig. 5 The detour from 0111 to 0000 (the nodes 0010, 0100, 0101, and 0110 are omitted)

not exist. However, if the number of node is greater than 8, the node labeled 8 exists and the detour $7_{(10)} \rightarrow 8_{(10)} \rightarrow 8_{(10)} \rightarrow 8_{(10)} \rightarrow 0$ (Fig. 5). This detour is based on the same direction vector but the applying order of levels in the routing table is changed.

Then, the following condition is added to detect the detour. Here, L_{max} is the maximum label in the routing table of the start node.

- $0 \leq a_{i+1} + b_{i-1} \times 2^{i} \leq L_{max}$ or $0 \leq a_{i+1} - b_{i-1} \times 2^{j+1} \leq L_{max}$

Furthermore, since two or more detour parts can exist in one direction vector, repeating to check of the above mentioned conditions in the direction vector is necessary.

Here, we consider the amount of reduction on the number of hops of the message forwarding by the detour. Let $sl = i - j$ in the above detour.. By the detection method of the detour, the amount of reduction is $sl - 3$. Therefore, if $sl > 3$ (i.e. the direction vector includes $\langle 0, 1, 1, 1 \rangle$ or $\langle 0, -1, -1, -1 \rangle$) then the detour is the shorter route than the initial route. In real situation, if the number of nodes is large and the level of Skip Graph is high then the possibility of the existence of the detour detected by our proposed method is not small. Then the amount of reduction is not small. Furthermore, the upper limit of the amount of reduction is $n - 2$. It is caused in case that $b_n = 0$ and the other b_i's are not 0 in the initial direction vector [i.e. $(0, 1, 1, \ldots, 1, 1)$ or $(0, -1, -1, \ldots, -1)$].

Our proposed detection method cannot always detect the shortest detour. But detecting the shortest detour by checking all cases is a very high cost operation and some combinational optimization is necessary. Our proposed detection method is very simple and its cost is very low. Then in real situation, it is expected that this method is effective.

3.3 Routing Algorithm for the Detour

In original Skip Graph, the route from the start node to the target node is decided by only key of the query condition. But in order to use the detour that is detected in Sect. 3.2, the routing algorithm is necessary to be changed as following.

1. Calculate the direction vector in the start node.
2. Calculate the detour and change the direction vector in the start node.
3. Select the leftmost non-zero element of the direction vector what the corresponding entry of the routing table is valid. Then, the destination node of the forwarding is decided to this element.
4. Forwarding to the destination node and change the corresponding element of the direction vector to 0.
5. Till reaching to the target node, repeat 3 and 4 in the node that received the forwarded message.

4 Experiments

To show effectiveness of our proposed routing algorithm, we evaluate the reduction of the number of hops in the query forwarding and compare the proposed routing algorithm to that of [5] and [13].

4.1 Experiment Environment

Since we evaluate only the number of hops of the query forwarding, the number of hops is calculated on a simulator. The bit length of key is 32 and exact retrievals and range retrievals are the target for the evaluation. In both the exact query and the range query, the number of hops of the message forwarding is calculated in all combination of the start node, the query condition. However, in range query, the range of the query condition is selected only from $\{2^r | 0 \leq r \leq 32\}$. By changing the number of nodes (from 10 to 10,000), the performance of the proposed routing algorithm is evaluated.

4.2 Results of Experiments

Figure 6 shows the average of the number of hops in exact retrieval, and Figs. 7, 8, and 9 are the average of the number of maximum hops in range retrieval. In these figures, *N* is the number of nodes, OPLH1 represents [13] using *N*, RKSG is the

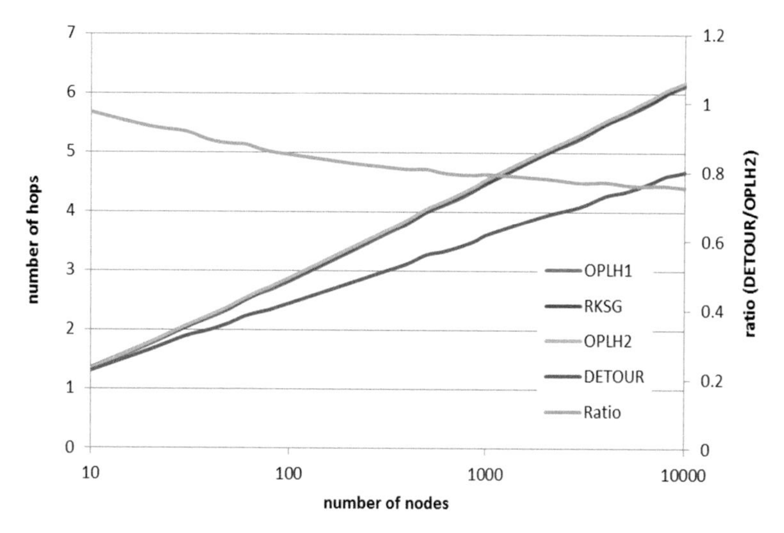

Fig. 6 The average of the number of hops for the exact retrieval

Range Key Skip Graph [5], and OPLH2 represents [13] not using N, and DETOUR represent our proposed routing algorithm. Ratio is the ratio DETOUR/OPLH2.

In Fig. 6, DETOUR is the best one. Furthermore, the more the number of nodes is, the more effective DETOUR becomes on Ratio. By increasing of the number of nodes, the level of Skip Graph becomes high, that is, the length of the direction vector becomes long. Then, the possibility of the existence of the detours becomes high. Moreover, the possibility that the amount of reduction is large becomes high because the upper limit of the amount of reduction becomes high. Therefore, the amount of reduction becomes large.

In Figs. 7, 8, and 9, DETOUR is also the best one. In these cases, the wider the range of the query condition is, the more effective DETOUR becomes. If the range

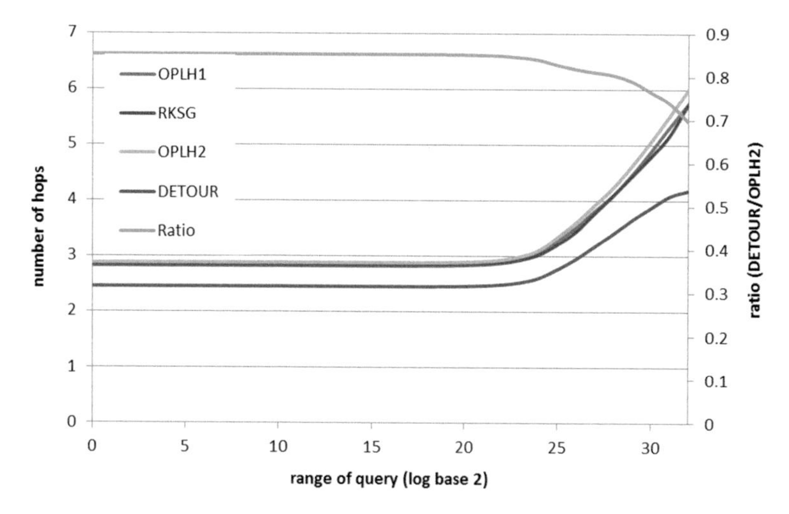

Fig. 7 The average of the number of maximum hops at range retrieval (N = 100)

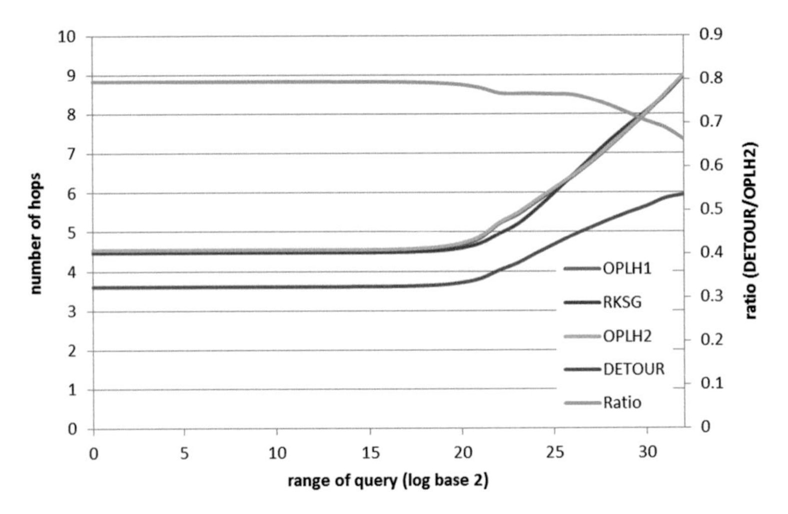

Fig. 8 The average of the number of maximum hops at range retrieval ($N = 100$)

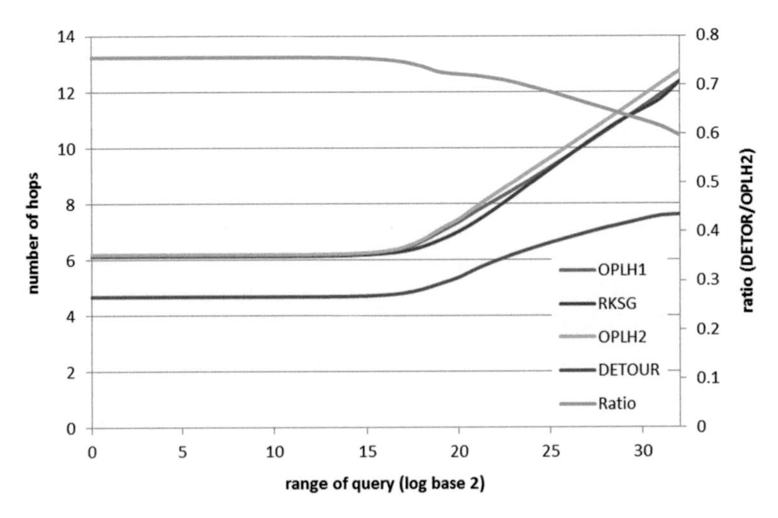

Fig. 9 The average of the number of maximum hops at range retrieval ($N = 10{,}000$)

of the query condition is wide then the number of the target node becomes large and the possibility of the existence of the detour becomes high. Then many detours are detected by the proposed method and the amount of the reduction is large. In all case, if the range of the query condition is narrow, the number of hops is fixed and the turning point exists. It is because the number of the target nodes is almost 1 when the range of the query condition is less than the range of the data in the node (2^{32-i} or 2^{32-i+1} where i is the hash level).

By experimental result, it can be said that the proposed routing algorithm is simple but it is effective.

5 Conclusions

We proposed a new routing algorithm for the distributed KVS based on OPLH and Skip Graph. Our algorithm detects detours and by using these detours, the number of hops of the message forwarding can be reduced. By experimental results, efficiency of our proposed algorithm is proved. For future works, we need to plan a routing algorithm suitable for replication.

References

1. Aspnes, J., Shah, G.: Skip graphs. ACM Trans. Algorithms **3**(4):37:1–37:25 (2007)
2. DeCandia, G., Hastorun, D., Jampani, M., Kakulapati, G., Lakshman, A., Pilchin, A., Sivasubramanian, S., Vosshal, P., Vogels, W.: Dynamo : Amazon's highly available key-value store. ACM SIGOPS Oper. Syst. Rev. **41**(6), 205–220 (2007)
3. Higuchi, K., Tsuji, T.: A distributed linear hashing enabling efficient retrieval for range queries. Proc. IEEE SMC **2010**, 838–842 (2010)
4. Ishi, Y., Teranishi, Y., Yoshida, M., Takeuchi, S., Shimojo, S., Nishio, S.: Range-key extensions of the skip graph. In: Proceedings of IEEE 2010 Global Communications Conference (IEEE GLOBECOM 2010), pp. 1–6 (2010)
5. Ishi, Y., Teranishi, Y., Yoshida, M., Takeuchi, S.: An implementation of large scale distributed key-value store with range search feature based on range-key skip graph. J. IPSJ **53** (7), 1850–1862 (2012)
6. Karger, D.R., Lehman, E., Leighton, F.T., Panigrahy, R., Levine, M.S., Lewin, D.: Consistent hashing and random trees: distributed caching protocols for relieving hot spots on the World Wide Web. In: ACM Symposium on Theory of Computing, pp. 654–663 (1997)
7. Konishi, Y., Yoshida, M., Takeuchi, S., Teranishi, Y., Harumoto, K., Shimojo, S.: An extension of skip graph to store multiple keys on single node. J. IPSJ **49**(9), 3223–3233 (2008)
8. Lakshman, A., Malik, P.: Cassandra—a decentralized structured storage system. ACM SIGOPS Oper. Syst. Rev. **42**(2), 35–40 (2010)
9. Litwin, W.: Linear hashing: new tool for file and table addressing. In: Proceedings of 6th Conference on Very Large DataBases, pp. 212–223 (1980)
10. Maymounkov, P., Mazieres, D.: Kademlia: a peer-to-peer information system based on the XOR metric. In: 1st International Workshop on Peer Systems (IPTPS'02), pp. 53–65 (2002)
11. Robinson, J.T.: Order preserving linear hashing using dynamic key statics. In: Proceedings of the 5th ACM SIGACT-SIGMOD symposium on principles of database systems, pp. 91–99 (1985)
12. Stoica, I., Morris, R., Karger, D., Kaashoek, F., Balakrishnan, H.: Chord: a scalable peer-to-peer lookup service for internet applications. ACM SIGCOMM **31**(4), 149–160 (2001)
13. Yoshida, M., Higuchi, K., Tsuji, T.: An implementation and evaluation of distributed key-value store based on order preserving linear hashing and skip graph. J. IEICE **J98-D**(5), 742–750

Effects of Lower Limb Cooling on the Work Performance and Physiological Responses During Maximal Endurance Exercise in Humans

Keiko Inoue, Masashi Kume and Tetsuya Yoshida

Abstract The purpose of this study was to investigate the effects of lower limb cooling on the work performance and physiological responses during maximal endurance exercise in humans. Eight male subjects underwent a maximal aerobic test using graded exercise on a cycle ergometer. The subjects wore trousers lined with tubes perfused water at 6 or 32 °C, and the target thigh muscle temperatures were 32 or 36 °C, respectively. The maximal working time was significantly lower during 32 °C than under 36 °C conditions. However, the body temperature, heat storage, heart rate and the total sweat loss were significantly lower under the 32 °C condition compared to those under the 36 °C condition. These results suggest that cooling the lower limbs to reach a thigh temperature of approximately 32 °C can reduce the physiological strain during maximal endurance exercise, although the endurance work performance under the 32 °C condition is lower than that under the 36 °C thigh temperature condition.

Keywords Muscle temperature · Maximal aerobic test · Cycle ergometer · Heart rate · Total sweat loss

K. Inoue · T. Yoshida (✉)
Kyoto Institute of Technology, Matsugasaki, Sakyo-Ku, Kyoto 606-8585, Japan
e-mail: yoshida@kit.jp

K. Inoue
e-mail: keiko5287@ab.auone-net.jp

M. Kume
Kyoto Bunkyo Junior College, Makishima, Senzoku80, Uji, Kyoto 611-0041, Japan
e-mail: kume@po.kbu.ac.jp

R. Lee (ed.), *Applied Computing & Information Technology*,
Studies in Computational Intelligence 619,
DOI 10.1007/978-3-319-26396-0_11

1 Introduction

In Japan, there are many traditional crafts and traditional industries which require a considerable amount of time and experience to become skillful [1–3]. Most traditional crafts involve making things by hand, which involves the use of various muscles in the body.

The environmental or tissue temperature affects the muscle activation [4, 5], where a high environmental temperature increases the heat stress in the body [6], causing a decrease in human work performance. Therefore, attenuating the heat stress can help to maintain the performance time to exhaustion [7], and the effects of pre-cooling on the body temperature and exercise performance have previously been investigated [8–10]. On the other hand, a low environmental temperature is related to a decreased tissue temperature, which attenuates the muscle contraction [11, 12]. However, the optimum muscle temperature that can attenuate physiological strain and maintain work performance is currently unclear.

We recently investigated the optimum muscle temperature required to maintain work performance, and found that a thigh temperature of approximately 36 °C may be optimal to maintain repeated maximal cycling exercise performance with the attenuation of heat stress [13]. An optimal muscle temperature would be expected to exist for performing endurance tasks, and we hypothesized that an extremely low active tissue temperature may attenuate the work performance, although the physiological strain at a low tissue temperature may be lower than that at a normal tissue temperature during the endurance task.

The goal of the present study was to determine the optimum muscle temperature required to maintain the endurance work performance with an attenuation of the physiological strain. As the first step to attain this goal, the present experiment utilized two muscle temperature conditions to test our hypothesis. The subjects in this study wore trousers lined with tubes perfused with water at 6 or 32 °C, to attain target thigh muscle temperatures of 32 or 36 °C, respectively. The subjects performed a maximal aerobic test using graded exercise on a cycle ergometer, and their work performance and physiological responses were measured during exercise to assess the effects of the different temperatures.

2 Method

2.1 Subjects

Eight college-aged male volunteers participated in this study with the approval of the Institutional Human Subjects Ethical Committee and after providing written informed consent. Their physical characteristics (mean ± SE) were an age of 24 ± 0.5 years old, height of 175 ± 2 cm, body mass of 64.8 ± 2.0 kg, thigh skinfold thickness of 5.5 ± 0.5 mm measured using a caliper-type adipometer (MK-60;

Yagami Inc., Tokyo, Japan) and a body fat percentage of 13.0 ± 0.8 % calculated from the formula described by Nagamine and Suzuki [14] using the skinfold thickness from the triceps and subscapula measured using an adipometer.

2.2 Experimental Procedure

On the day of the experiment, each subject reported to the laboratory without having eaten lunch, and then drank 200–300 ml of water between 12 noon and 1 PM to avoid dehydration. To help control for the physiological condition for subjects, they had refrained from heavy exercise for 24 h and also from the intake of salty food, alcohol and caffeine for 17 h before arriving at the laboratory. After they drank the water, the nude body mass was measured using a scale with a 10 g precision (FW-100 K, A&D, Japan). Each subject dressed in a long-sleeved T-shirt (XA6045, Asics, Tokyo, Japan) and swimming trunks, and wore tube-lined trousers (Microclimate systems, Delta Temax Inc, Ontario, Canada, Fig. 1) that were used to cool the lower limbs, including the hips and abdomen, by water perfusion. They all had instruments put into place to carry out the physiological measurements. After a 30-min resting period, water perfusion into the tube-lined trousers was started. The temperature of the water (6 or 32 °C) that was perfused through the tube-lined trousers was controlled using a low temperature water bath (LTB-400, Iuchi,

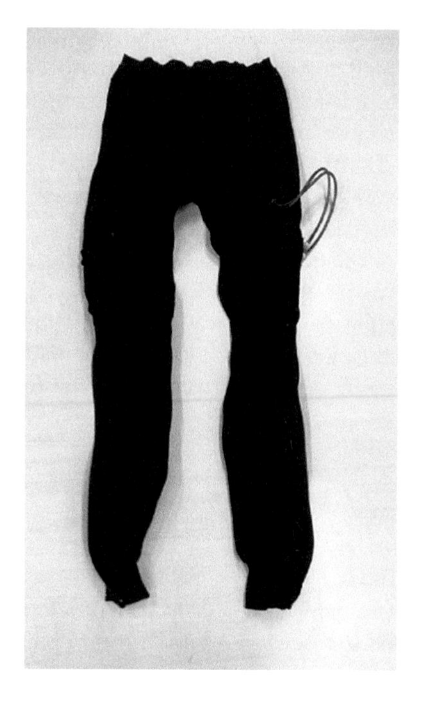

Fig. 1 The tube-lined trousers used in this experiment

Fig. 2 The maximal aerobic test performed using a cycle ergometer, and the measurement of expired air in a subject

Japan). The water flow rate was set at 0.6 L/min using a seal-less canned motor pump (UPS 25-8- JA, Grundfos, Denmark). The subject sat down on a chair and rested until the deep thigh temperature reached the targeted value (32 or 36 °C) in a room maintained at a thermo-neutral environment (an ambient temperature of 24.1 ± 0.2 °C; relative humidity of 55 ± 3 %). After reaching the target temperature or close to the target, the subject moved to a bicycle ergometer (POWERMAX-VII, CONBI, Japan) and commenced exercise at a 60 rpm pedaling rate, and the workload was gradually increased by 0.3 kp/min until the subjects voluntarily became exhausted (Fig. 2).

The graded cycling exercise experiments were performed twice: one experiment was performed under the 32 °C thigh temperature with water perfusion at 6 °C, while the other was performed under the 36 °C thigh temperature with water perfusion at 32 °C, with an interval of at least one week between the experiments, which were performed in random order.

2.3 *Measurements*

The deep thigh temperature (Tdt), as an index of the active tissue temperature, was measured using a deep body thermometer (Coretemp CM-210, Terumo, Japan) by the zero-heat-flow (ZHF) method. The temperature of the deep body tissue measured by this method has been reported to have a high correlation with the muscle temperature at a depth of 18 mm [15]. The sensor elements (CM-210 PD1, Terumo, Japan), which were 4.3 cm in diameter, were fixed securely with tape and were placed over the subject's right vastus lateralis, 10 cm superior to the top of the patella. The sensor of the thermometer was covered with an insulation material so that the tubes of the trousers did not touch it. The measurement of the Tdt was performed every 30 s during the experiments. The tympanic temperature (Tty) was measured using an infrared tympanic thermometer (Genius, Argyle, Japan) and was recorded before water perfusion into the tube-lined trousers, at the start of exercise, after four and 8 min of exercise and at the end of exercise. The skin temperature was measured every 30 s at three sites (chest, upper-arm and thigh) with thermocouple probes, and the mean skin temperature (Tsk) was calculated from the equation presented by Roberts et al. [16]; $0.43 \times$ chest + $0.32 \times$ thigh + $0.25 \times$ upper-arm. The mean body temperature (Tb) was calculated from the Tty and Tsk using the equation: $Tb = 0.9 \times Tty + 0.1 \times Tsk$ [17], and the heat storage (S) was also calculated as: body weight × increase in Tb (ΔTb) $\times 3474 \times s^{-1}$ W [18]. Before every experiment, the thermocouple probes were calibrated by the immersion of each probe into water at 35–37 °C.

Expired air was collected, and the O_2/CO_2 levels were analyzed via a computer on-line system (METS-900, VISEMEDICAL, Japan). To ensure the attainment of a valid peak oxygen uptake (VO_{2peak}), as an index of aerobic capacity, the following two criteria were met by each of the subjects: (1) a plateau in O_2 consumption with an increasing work load, (2) a respiratory exchange ratio of at least 1.15 at the end of exercise. The maximal working time was defined as the time at the end of exercise when subjects became voluntarily exhausted, but not at the time when the VO_{2peak} was observed. The total sweat loss (TSL) was determined from the nude body weight obtained before and after the experiment, and it was corrected for the respiratory water loss [19]. The heart rate was measured using a heart rate monitor (RS400, Pola, Finland) and was recorded every one minute beginning before water perfusion.

2.4 Statistical Analyses

All variables measured during the experiments are shown as the means ± standard error (SE). To compare the two tissue temperature conditions, a two-way repeated measures ANOVA (condition × time) was first performed, and the Bonferroni test was used for the post hoc analyses when significant interactions were observed. The significance levels of all tests were set at <5, <1 and <0.1 %.

3 Results

3.1 Body Temperature and Heat Storage (S)

Figures 3 and 4 show the changes in the Tdt and Tty, respectively, during the experiment. The Tdt values just before exercise under the 32 and 36 °C conditions were 32.1 ± 0.26 and 36.1 ± 0.07 °C, respectively. The Tdt under both experimental conditions increased at the end of exercise. However, the increase in the Tdt under the 32 °C condition was greater than that under the 36 °C condition. The Tty during the water perfusion before exercise was similar under the two temperature conditions. Because of the small changes in the Tty observed during graded cycling exercise, the increase in the Tty (ΔTty) from baseline (before perfusion) was shown in Fig. 4. The Tty under the 32 °C condition decreased until the end of exercise ($p < 0.001$), however, that under the 36 °C condition increased until the end of exercise ($p < 0.05$). Figures 5 and 6 show the changes in the Tsk and Tb, respectively, during the experiment. The Tsk and Tb were significantly lower under the 32 °C condition than under the 36 °C condition, and both parameters had significantly increased at the end of exercise under the 36 °C condition. The values of S from before perfusion to before exercise, from the start to the end of exercise, and from before perfusion to the end of exercise under the 32 and 36 °C conditions are shown in Fig. 7. The S values for all three phases were significantly lower under the 32 °C condition than under the 36 °C condition.

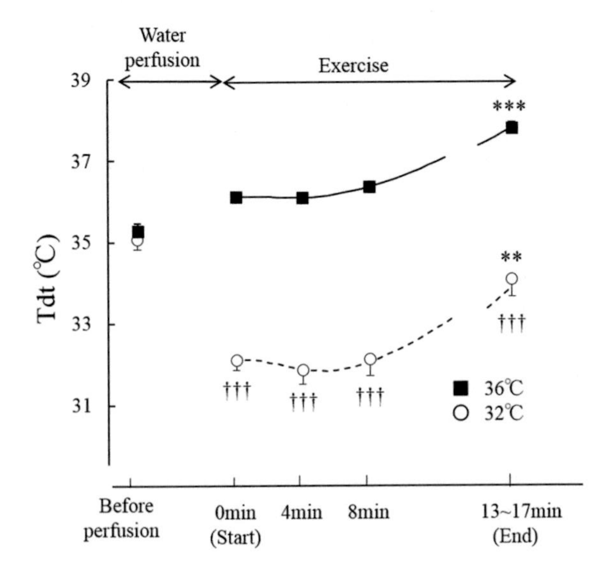

Fig. 3 The changes in the deep thigh temperature (Tdt) from "Before water perfusion" to the "End" under the 32 and 36 °C conditions. The data are expressed as the means ± SE. **($p < 0.01$) and ***($p < 0.001$) show significant differences from the "Start" level; †††($p < 0.001$) shows a significant difference between conditions

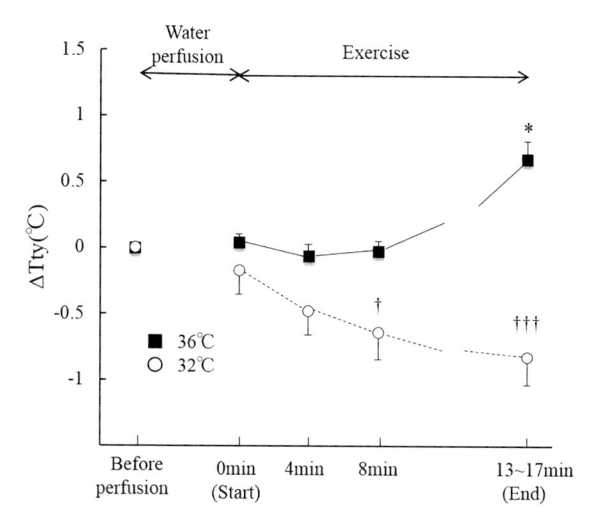

Fig. 4 The changes in the increase in the tympanic temperature (ΔTty) from "Before perfusion" to the "End" under the 32 and 36 °C conditions. The data are expressed as the means ± SE. * ($p < 0.05$) shows a significant difference from the "Start" level; †($p < 0.05$) and †††($p < 0.001$) show significant differences between conditions

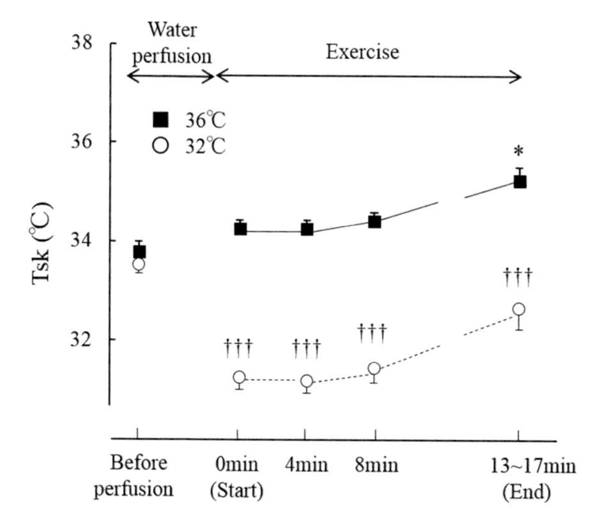

Fig. 5 The changes in the mean skin temperature (Tsk) from "Before perfusion" to the "End" under the 32 and 36 °C conditions. The data are expressed as the means ± SE. *($p < 0.05$) shows a significant difference from the "Start" level; †††($p < 0.001$) shows a significant difference between conditions

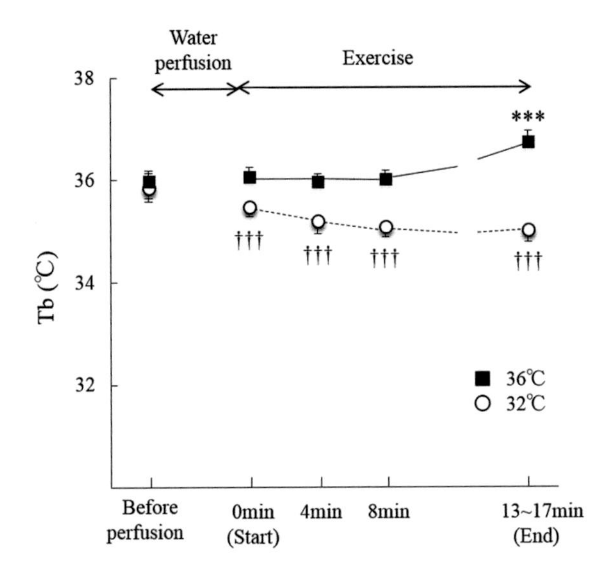

Fig. 6 The changes in the mean body temperature (Tb) from "Before perfusion" to the "End" under the 32 and 36 °C conditions. The data are expressed as the means ± SE. ***($p < 0.001$) shows a significant difference from the "Start" level; †††($p < 0.001$) shows a significant difference between conditions

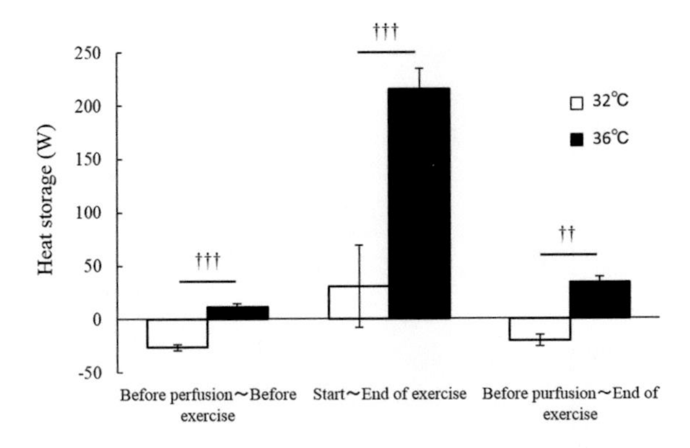

Fig. 7 A comparison of the heat storage under the 32 and 36 °C conditions during three phases (from the before perfusion to before exercise, from the start to the end of exercise, and from the before perfusion to the end of exercise). The data are expressed as the means ± SE. ††($p < 0.01$) and †††($p < 0.001$) show significant differences between conditions

3.2 Physiological Responses

A comparison of the oxygen uptake (VO_2) and heart rate under the 32 and 36 °C conditions is shown in Figs. 8 and 9, respectively. The VO_2 increased under both the 32 and 36 °C conditions according to the increases in workload. There were no

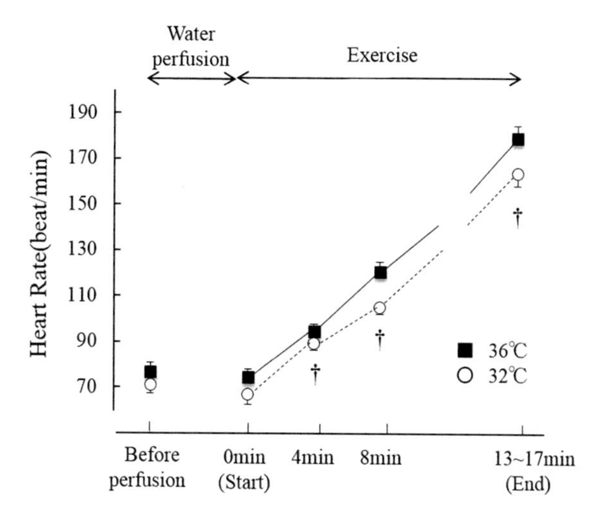

Fig. 8 The changes in the heart rates from "Before perfusion" to the "End" under the 32 and 36 °C conditions. The data are expressed as the means ± SE. †($p < 0.05$) shows a significant difference between conditions

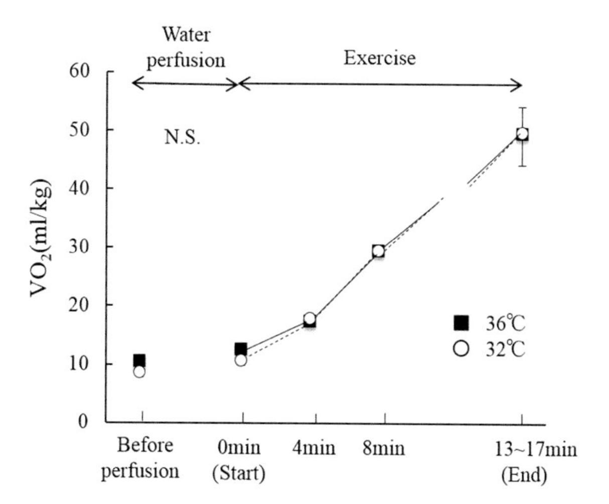

Fig. 9 The changes in the oxygen uptake (VO_2) from "Before perfusion" to the "End" under the 32 and 36 °C conditions. The data are expressed as the means ± SE. *N.S.* not significant

significant differences in the oxygen uptake between the 32 and 36 °C conditions; however, the heart rate was significantly lower under the 32 °C condition than under the 36 °C condition. The TSL was also significantly lower under the 32 °C condition (0.06 kg ± 0.004 kg/hour) than under the 36 °C condition (0.23 kg ± 0.03 kg/hour).

3.3 Work Performance

The mean peak oxygen uptake (VO_{2peak}) and mean maximal working time under the 32 and 36 °C conditions were 50.4 (rang 35.3–71.7) ml/kg/min and 15 min

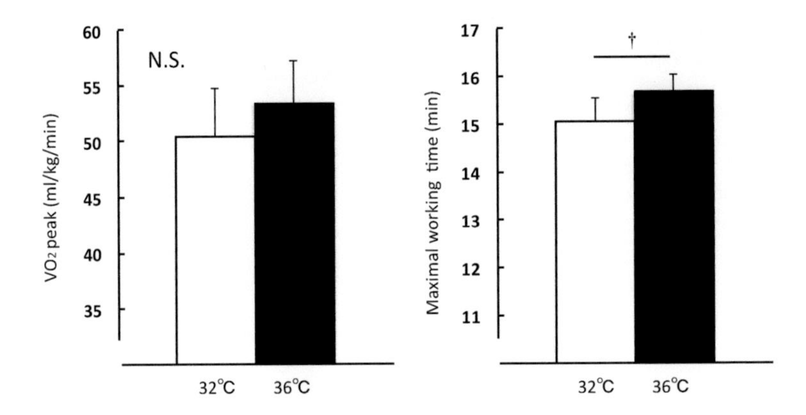

Fig. 10 A comparison of the peak oxygen uptake (VO_{2peak}, *left graph*) and maximal working time (*right graph*) during the 32 and 36 °C conditions. The data are expressed as the means ± SE. †($p < 0.05$) shows a significant difference between conditions; *N.S.* not significant

(range 13–17) min and 53.4 (35.5–66.3) ml/kg/min and 16 (range 14–17) min, respectively. As shown in Fig. 10, the VO_{2peak} tended to be lower and the maximal working time was significantly lower under the 32 °C condition compared to the 36 °C condition.

4 Discussion

This is the first study to examine the effects of the muscle temperature on the physiological strain and work performance during maximal endurance cycling exercise using tube-lined trousers. The new finding in the present study is that cooling the lower limbs to a thigh temperature of approximately 32 °C can reduce the physiological strain during maximal endurance exercise, although the maximal working time under 32 °C condition is lower than that under the 36 °C thigh temperature condition.

4.1 Body Temperature

In the present study, although the Tdt, Tsk and Tb were significantly lower under the 32 °C condition than under the 36 °C condition, the Tty at the start of exercise was similar between the conditions. During body cooling, the skin blood flow decreases and the heat production caused by shivering and non-shivering movements increases to attenuate the falling body temperature [20]. At the end of exercise, however, the Tty was significantly elevated under the 36 °C condition, but was decreased under the 32 °C condition compared to the baseline. Because of the increase in blood circulation during exercise, the blood returning to the heart from the skin with the low (32 °C) temperature might have caused a decrease in the Tty under the 32 °C condition. Furthermore, the S was significantly greater under the 36 °C condition than under the 32 °C condition, suggesting that heat production greater than the heat loss during exercise might have led to both a greater S and an increase in the Tty under the 36 °C condition.

4.2 Physiological Response

It is well known that elevation of the body temperature leads to physiological stress [6, 7, 18], while body cooling attenuates the physiological strain [8–10]. In the present study, the HR and TSL were greater under the 36 °C condition than under the 32 °C condition. Because of the greater S and increase in the Tty, an enhancement of the physiological strain might be observed under the 36 °C condition during maximal endurance exercise. In other words, cooling the lower limbs to reach a thigh temperature of approximately 32 °C could reduce the physiological strain during maximal endurance exercise.

4.3 Work Performance

An increase in muscle temperature has been reported to increase the oxygen release from hemoglobin and myoglobin [21–23], the transmission rate of nerve impulses [24], glycogenolysis, glycolysis and high-energy phosphate degradation [25–28], and to decrease the stiffness of muscles and joints [29, 30]. Thus, an increase in the muscle temperature has the potential to improve the exercise performance. In other words, low muscle temperatures might lead to reduce exercise performance. In the present study, although the VO_{2peak} tended to be lower and the maximal working time was significantly lower under the 32 °C condition than under the 36 °C condition, the VO_2 from the start to the end of exercise was similar between conditions. These results suggest that the oxygen utilization capacity did not differ between the 32 and 36 °C conditions during sub-maximal exercise with an increase in workload. In the present study, however, it was demonstrated that cooling the lower limbs to a thigh temperature of approximately 32 °C could reduce the maximal endurance capacity. Thus, the reduction of the maximal endurance capacity due to lower limb cooling might be attributable to a decrease in the physiological responses, with the exception of a decrease in the oxygen utilization capacity.

5 Conclusion

The present study demonstrated that cooling the lower limbs could reduce the physiological strain during maximal endurance exercise, although the maximal working time under the cooling condition was lower than under the normal muscle temperature condition. Thus, an optimal muscle temperature should exist for performing endurance tasks with attenuation of physiological strain, and optimal environments or tissue temperatures would also exist for exhibiting the best skills and techniques of traditional craft workers. Future experiments should be performed to evaluate the performance of traditional skills under optimal environmental or tissue temperature conditions.

References

1. Chen, G., Yamashita, A., Shibata, K., et al.: Three-dimensional analysis of the work motion using woodworking tools 1. J. Wood Sci. **48**, 80–88 (2002)
2. Yamaguchi, T., Kitamura, K., Uenishi, T., et al.: Comparative analysis of skilled and unskilled behaviors in sewing-machine operation. Biomechanism **16**, 207–220 (2002)
3. Yoshida, T., Ohnishi, A., Shirato, M., et al.: Characteristics of "TAKUMINO-WAZA" in Japanese traditional craft. Mater. Integr. **21**, 20–25 (2008)
4. Nybo, L., Jensen, T., Nielsen, B., et al.: Effects of marked hyperthermia with and without dehydration on VO_2 kinetics during intense exercise. J. Appl. Physiol. **90**, 1057–1064 (2001)
5. Racinais, S., Oksa, J.: Temperature and neuromuscular function. Scand. J. Med. Sci. Sports **20** (Suppl. 3), 1–18 (2010)
6. Gonzalez-Alonso, J., Teller, C., Andersen, S.L., et al.: Infuluence of body temperature on the development of fatigue during prolonged exercise in the heat. J. Appl. Physiol. **86**, 1032–1039 (1999)
7. Nielsen, B., Hales, J.R.S., Strange, S., et al.: Human circulatory and thermoregulatory adaptations with heat acclimation and exercise in a hot, dry environment. J. Physiol. **460**, 467–485 (1993)
8. Booth, J., Marino, F., Ward, J.J.: Improved running performance in hot humid conditions following whole body precooling. Med. Sci. Sports Exerc. **29**, 943–949 (1997)
9. Lee, D.T., Haymes, E.M.: Exercise duration and thermoregulatory responses after whole body precooling. J. Appl. Physiol. **79**, 1971–1976 (1995)
10. Marino, F.E.: Methods, advantages, and limitations of body cooling for exercise performance. Br. J. Sports Med. **36**, 89–94 (2002)
11. Ball, D., Burrows, C., Sargeant, A.J.: Human power output during repeated sprint exercise. Eur. J. Appl. Physiol. **79**, 360–366 (1999)
12. Oksa, J., Rintamaki, H., Rissanen, S.: Muscle performance and electromyogram activity of the lower leg muscles with different levels of cold exposure. Eur. J. Appl. Physiol. **75**, 484–490 (1997)
13. Inoue, K., Kume, M., Yoshida, T.: Determination of the optimum muscle temperature for maintaining work performance with attenuation of heat stress in humans. In: 2014 IIAI 3rd International Conference on Advanced Applied Informatics, pp. 795–800 (2014)
14. Nagamine, S., Suzuki, S.: Anthropometry and body composition of Japanese young men and women. Human Biol. **36**, 8–15 (1964)

15. Matsukawa, T., Kashimoto, S., Ozaki, M., et al.: Temperatures measured by a deep body thermometer (Coretemp) compard with tissue temperatures measured at various depths using needles placed into the sole of the foot. Eur. J. Anaesthesiol. **13**, 340–345 (1996)
16. Roberts, M.F., Wenger, C.B., Stolwijk, J.A., et al.: Skin blood flow and sweating changes following exercise training and heat acclimation. J. Appl. Physiol. **43**, 133–137 (1977)
17. Gagge, A.P., Nishi, Y.: Heat exchange between human skin surface and thermal environment. In: Lee, D.H.K. (ed.) Handbook of physiology. American Physiological Society, Bethesda (1977)
18. Nielsen, B.: Solar heat load: heat balance during exercise in clothed subjects. Eur. J. Appl. Physiol. **60**, 15–25 (1990)
19. Mitchell, J.W., Nadel, E.R., Stolwijk, J.A.J.: Respiratory weight loss during exercise. J. Appl. Physiol. **32**, 474–476 (1972)
20. Yanagisawa, O., Homma, T., Okuwaki, T., et al.: Effects of cooling on human skin and skeletal muscle. Eur. J. Appl. Physiol. **100**, 737–745 (2007)
21. Barcroft, J., King, W.O.R.: The effect of temperature on the dissociation curve of blood. J. Physiol. **39**, 374–384 (1909)
22. McCutcheon, L.J., Geor, R.J., Hinchcliff, K.W.: Effects of prior exercise on muscle metabolism during sprint exercise in humans. J. Appl. Physiol. **87**, 1914–1922 (1999)
23. Theorell, H.: The effect of temperature on myoglobin. Biochem. Z **73**, 268 (1934)
24. Karvonen, J.: Importance of warm up and cool down on exercise performance. In: Karvonen, J., Lemon, P.W.R., Iliev, I. (eds.) Medicine and sports training and coaching. Karger, Basel (1992)
25. Edwards, R.H.T., Harris, R.C., Hultman, E., et al.: Effect of temperature on muscle energy metabolism and endurance during successive isometric contractions, sustained to fatigue, of the quadriceps muscle in man. J. Physiol. **220**, 335–352 (1972)
26. Febbraio, M.A., Carey, M.F., Snow, R.J., et al.: Influence of elevated muscle temperature on metabolism during intense, dynamic exercise. Am. J. Physiol. **271**, R1251–R1255 (1996)
27. Febbraio, M.A.: Does muscle function and metabolism affect exercise performance in the heat? Exerc. Sport Sci. Rev. **28**, 171–176 (2000)
28. Hargeaves, M., Angus, D., Howlett, K., et al.: Effect of heat stress on glucose kinetics during exercise. J. Appl. Physiol. **81**, 1594–1597 (1996)
29. Buchthal, F., Kaiser, E., Knappeis, G.G.: Elasticity, viscosity and plasticity in the cross started muscle fibre [abstract]. Acta Physiol. Scand. **8**, 1637 (1994)
30. Wright, V., Johns, R.J.: Quantitative and qualitative analysis of joint stiffness in normal subjects and in patients with connective tissue disease. Ann. Rheum. Dis. **20**, 36–46 (1961)

Demonstration of Pedestrian Movement Support System Using Visible Light Communication

Saeko Oshiba, Shunsuke Iki, Hirotoshi Kii, Hiroki Watanabe, Yusuke Murata, Yuki Nagai, Jun Yamazaki, Masato Yoshihisa, Yoji Kitani, Saori Kitaguchi and Kazunari Morimoto

Abstract We constructed a pedestrian support system for the visually impaired that uses visible light communication (VLC) from self-illuminating bollards. To achieve this goal, we conducted a visibility evaluation using pedestrians with low and normal vision to determine the best bollard lighting pattern for pedestrian navigation systems. Furthermore, we designed a receiving terminal that is easy to manage and can be operated by touch alone. Moreover, we constructed a sidewalk model to evaluate the navigation system, and the communication area of the system was experimentally determined. We then created a prototype of the navigation system using VLC. The prototype was evaluated by visually impaired users, clarifying the effectiveness of the system.

Keywords Bollard · Pedestrian support · Visible-light communication · Visual impairment

1 Introduction

In Japan, currently, there are approximately 300 thousand people whose corrected eyesight in both eyes is less than 0.3. They can often find daily life difficult. Further, 70 % of these people need help from others and hardly go out [1]. The importance of social support for individuals with visual impairment has also become increasingly clear. Future social environments should consider visually impaired and blind people [2].

Global Positioning System (GPS) is commonly used for navigation. However, the accuracy of GPS walking assistance deteriorates in the shadow of buildings and in indoor environments.

S. Oshiba (✉) · S. Iki · H. Kii · H. Watanabe · Y. Murata · Y. Nagai · J. Yamazaki · M. Yoshihisa · Y. Kitani · S. Kitaguchi · K. Morimoto
Kyoto Institute of Technology, Matsugasaki, Sakyoku, Kyoto, Japan
e-mail: oshiba@kit.ac.jp

R. Lee (ed.), *Applied Computing & Information Technology*,
Studies in Computational Intelligence 619,
DOI 10.1007/978-3-319-26396-0_12

We propose a pedestrian movement support system using visible light communication (VLC) with light-emitting diodes (LEDs) [3, 4]. VLC is wireless communication that uses light in the human-visible wavelength range, from 380 to 780 nm, and is the only communication technology that can be incorporated into existing illumination [5]. The directionality of LEDs limits the communication area so that VLC can transmit detailed information. Furthermore, pedestrians can be guided with high accuracy by installing VLC in places where GPS cannot be used. In this study, we construct a VLC system using illuminated bollards. A bollard is a short post to prevent vehicles from entering certain areas. They indicate the boundary between a sidewalk and roadway.

In this paper, we report the following. First, we experimentally evaluated bollard light patterns for visually impaired people to determine the best light pattern. In addition, we designed a receiving terminal that is easy handle and manage. Moreover, we determined the communication area of our proposed sidewalk model. As a result, we developed a prototype of the navigation system using VLC. Finally, the prototype was evaluated by visually impaired users to determine the effectiveness of the system.

2 Sidewalk Model

We constructed a sidewalk model, shown in Fig. 1, to evaluate the navigation system. This model meets Kyoto road structure ordinances (i.e., bollards can be installed in the road where the sidewalks are 2.0 m or wider, and the effective sidewalk width is 1.75 m or more).

In this model, tactile tiles are installed that lead to a crosswalk with signals. Illuminated bollards are located 1.75 m from the tactile tiles and installed in intervals of 5 m along the sidewalk. The LEDs in the bollards face into the road at an angle of 45° from the boundary between the sidewalk and the roadway. The sidewalk is also lit by normal streetlights.

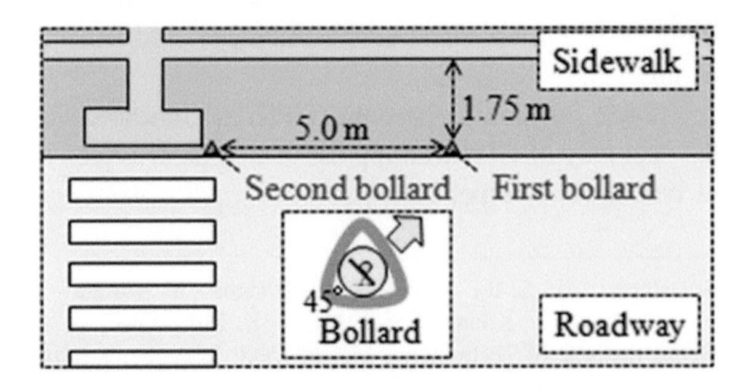

Fig. 1 Sidewalk model

3 Design of Receiving Terminal

We designed a receiving terminal, keeping in mind the following two points.

(1) Visually impaired people should be able to distinguish between the front and back as well as the top and bottom of the receiving terminal.
(2) The terminal should feel comfortable in the hand, easy to hold, and easy to carry.

The completed design is shown in Fig. 2. An image of its use is shown in Fig. 3.

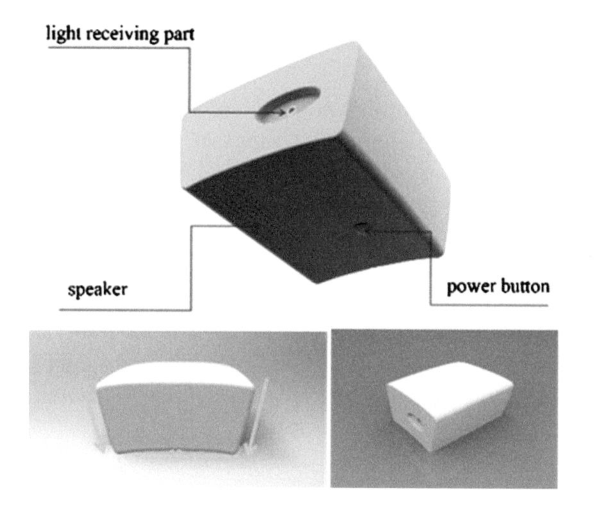

Fig. 2 Images of the receiving terminal

Fig. 3 Use of the receiving terminal

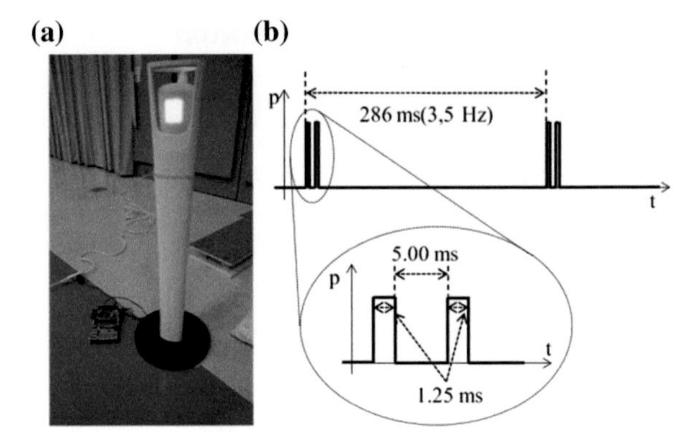

Fig. 4 Illuminated bollard. **a** Image. **b** Light emission patterns

4 VLC Design

4.1 Prototype

We constructed prototype versions of the VLC system. The VLC transmitter is installed in the illuminated bollard, as shown in Fig. 4a. The bollard is designed to be visible and durable [3], and it flashes at a rate of 3.5 Hz. The flash time width is about 7.5 ms (comprising two pulses separated by a 5 ms interval), as shown in Fig. 4b. The data signal is superimposed on the blinking light emission of the bollard. We transmit the data during one flash pulse (pulse width: 1.25 ms). The data rate is 125 kbps so that a sufficient volume of data can be transmitted within the time period of the pulse. Because pulse-position modulation (PPM) is a technique that achieves very good average-power efficiency, we used 4-PPM as the signal modulation of the VLC transmitter.

4.2 Receive Condition

In this system, visually impaired people aim the receiving terminal at the bollards in order to communicate with them. We investigated whether they could correctly aim the terminal. The experimental setup is shown in Fig. 5. The height of the receiving terminal is the same as the height of the bollards. Visually impaired people aimed the receiving terminal at the bollards, and we then measured its angle. In this setup, the perpendicular direction to the tactile tiles is at 0°. Five visually impaired people, including three men and two women, participated in the experiment. The bollard luminescent patterns are shown in Table 1. As the flash time-width increases, the LED becomes brighter. In this experiment, three bollard luminescent patterns, 1, 4, and 6, were tested.

The experimental results are shown in Fig. 6. The plot represents relationships between angles of the receiving terminal and bollard luminescent patterns. The

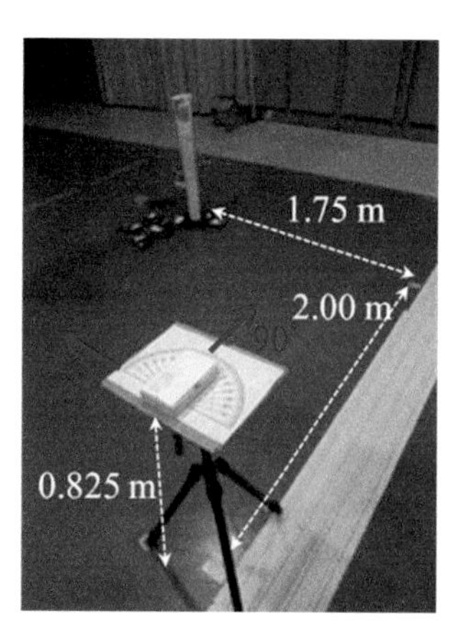

Fig. 5 Experimental setup

Table 1 Bollard luminescent patterns

No	Number of flashes	Data rate (kbps)	Number of commands	Flash time-width (ms)
1	1	4.8	8	29.952
2	1	4.8	3	11.232
3	2	125	8	2.304
4	1	125	8	1.152
5	1	125	3	0.432
6	1	250	3	0.108

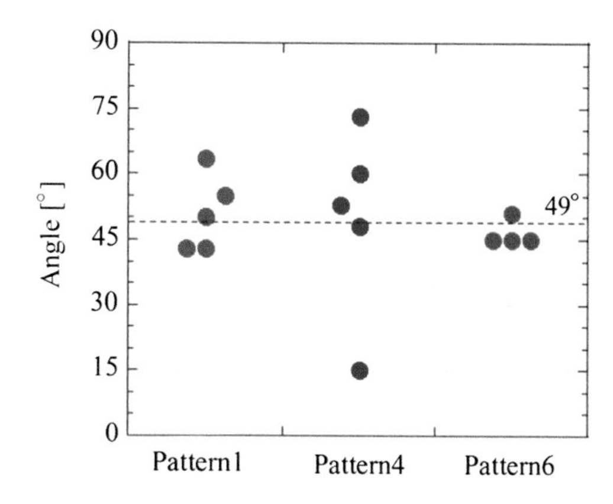

Fig. 6 Experimental result for receiver angle

dotted line in Fig. 6 indicates the direct angle from the receiving terminal to the bollard at 2.0 m distance (49°). The direction error increases for patterns 6, 1, and 4, in that order. Most participants were able to identify the bollard location more exactly using pattern 6, which is the darkest. However, one participant could not find the receiving terminal in pattern 6. Brighter patterns make it difficult to identify direction because light diffuses as it becomes brighter. Pattern 4's angle error is the greatest, although all participants could recognize the bollard light. The angle of the receiving terminal is 15°–75° and the maximum angle error is 34°.

4.3 Communication Area Determination

We measured the communication area of this prototype by measuring the receiver voltage when walking along the tactile tiles in the sidewalk model. The experimental setup is shown in Fig. 7, where x is distance and α is receiver angle. The receiving terminal was fixed at a height of 82.5 cm. In addition receiver voltages were measured while varying its angle α from 0° to 90°.

The results are shown in Fig. 8. In this figure, the horizontal axis represents the distance x from the bollard, and the vertical axis represents the received voltage. The receiver voltage-distance curves are almost same for receiver angles of 15°–75°. Given this result, we adjusted the threshold voltage of the receiving terminal to

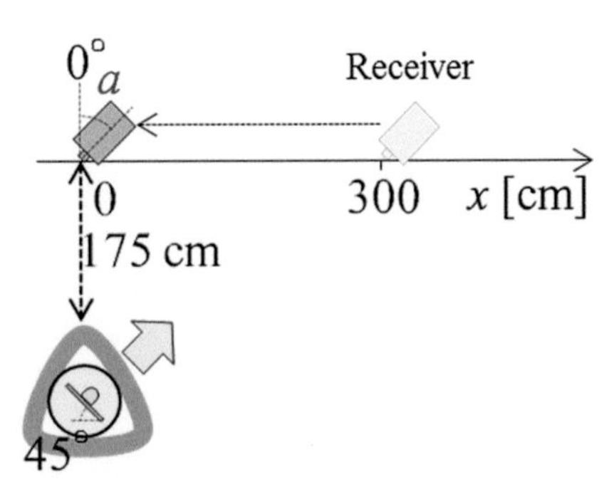

Fig. 7 Experimental setup

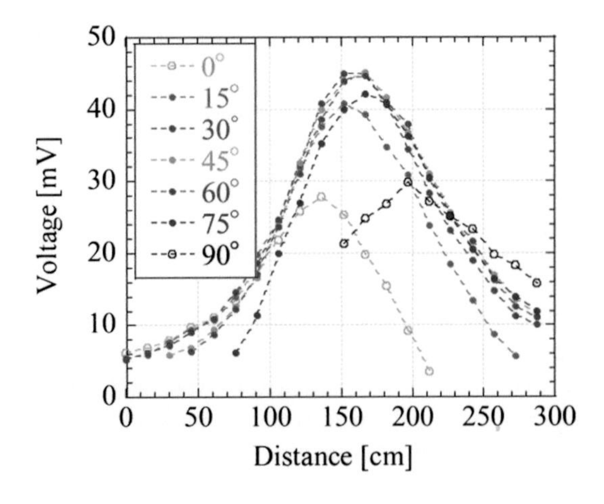

Fig. 8 Experimental result

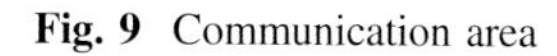

Fig. 9 Communication area

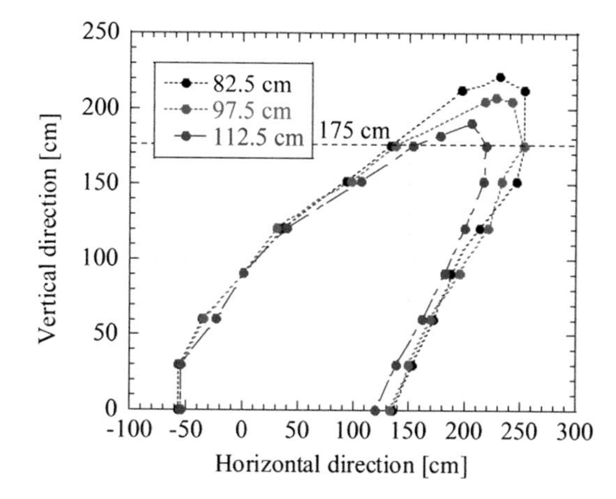

30 mV, and determined that signal was able to be received at distance of 120–200 m for angles from 15 to 75°.

We then measured the communication area, which is the area within which the signal can be received at a height of 82.5–112.5 cm when the receiving terminal is aimed towards the bollards. The experimental results are shown in Fig. 9.

In this figure, the horizontal and vertical axes are parallel and perpendicular, respectively, to the tactile tiles. The illuminated bollard is located at the origin. As shown in Fig. 9, at a vertical distance of 1.75 m, the signal can be received at horizontal distance of 150–200 m.

5 Experiments

We conducted a visibility evaluation using pedestrians with low and normal vision to determine the best bollard lighting pattern for pedestrian navigation systems.

5.1 Evaluation Using Visually Impaired Pedestrians

We conducted a visibility evaluation experiment with fourteen visually impaired people consisting of seven men and seven women. The experiment was performed in a room without outside light because the navigation system is assumed to operate at night.

The bollard luminescent patterns used in the experiment are same as those listed in Table 1. The luminescent colors of the bollard light in this experiment were white and yellow.

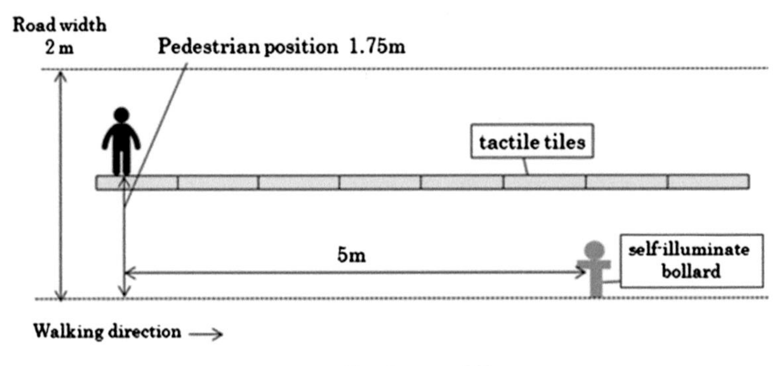

Fig. 10 Setup of evaluation experiment

Table 2 Evaluation method

	Very	Somewhat	Neutral	Somewhat	Very	
Glare	1	2	3	4	5	No glare
Bright	1	2	3	4	5	Dark
Not favorite	1	2	3	4	5	Favorite

Because visibility is important while walking, the subjects should be able to see the bollards while on foot. As shown in Fig. 10, the illuminated bollards are located 1.75 m from the tactile tiles. The subjects were instructed to walk on a tactile tile at a distance of 5 m from the illuminated bollards. They evaluated the visibility afterwards. The evaluation method is shown in Table 2. Five rankings were used to evaluate glare, brightness, and favorite pattern.

Figure 11 shows the experimental results. In the case of white, patterns 1–2 were "glare" and "flickering" and pattern 6 was "dark". In the case of yellow, pattern 3 4, 5, and 6 were "dark". The visually impaired people preferred patterns 3–5 in white, and patterns 1–2 in yellow.

5.2 *Evaluation Using Pedestrians with Normal Vision*

Participants with normal vision were used to conduct the same experiment as described in Sect. 5.1. These participants noted that patterns 1–2 in both white and yellow were "tiring" and "flickering". Therefore, to avoid irritating pedestrians with normal vision, patterns 1–2 were avoided.

The experiment found that both normal and low-vision users preferred the white patterns 3–5. Among these patterns, patterns 4 and 5 were less tiring to look at than pattern 3. In addition, if pattern communication characteristics are considered, pattern 4 is the best. In subsequent experiments in this study, the illuminated bollards used white pattern 4.

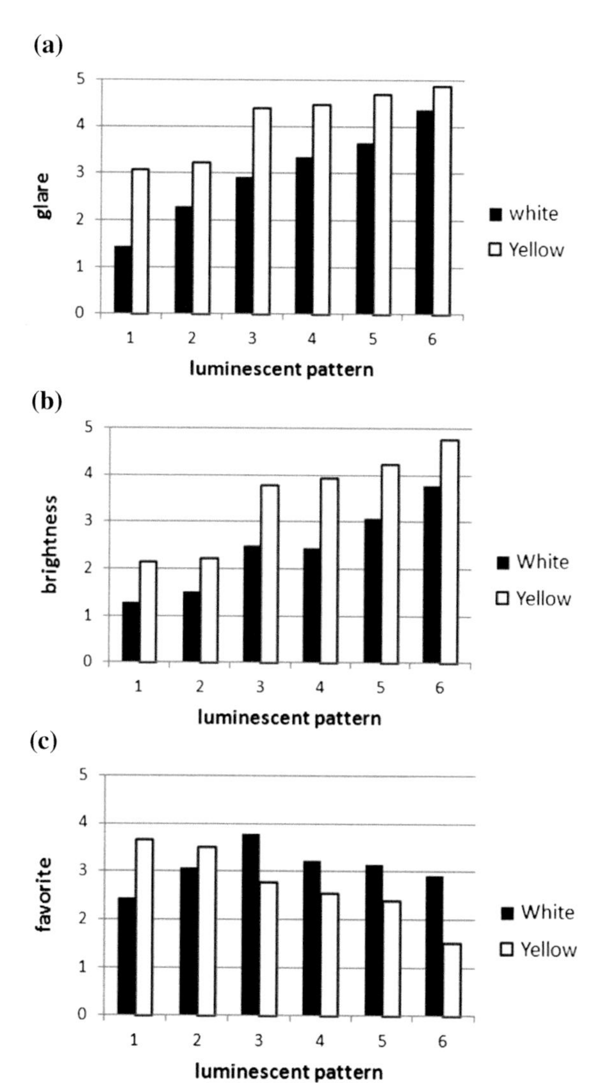

Fig. 11 Results of subjective evaluation. **a** Glare. **b** Brightness. **c** Favorite

6 Evaluation of the Prototype

6.1 Experimental Procedure

We developed a prototype model of the pedestrian movement support system using VLC and evaluated it experimentally. Six visually impaired people, one man and five women, participated in the evaluation. We performed the experiments on the sidewalk models shown in Fig. 12. Three street configurations with corresponding messages were tested.

Fig. 12 Sidewalk models for evaluation experiments. **a** Configuration 1. **b** Configuration 2. **c** Configuration 3

(a)

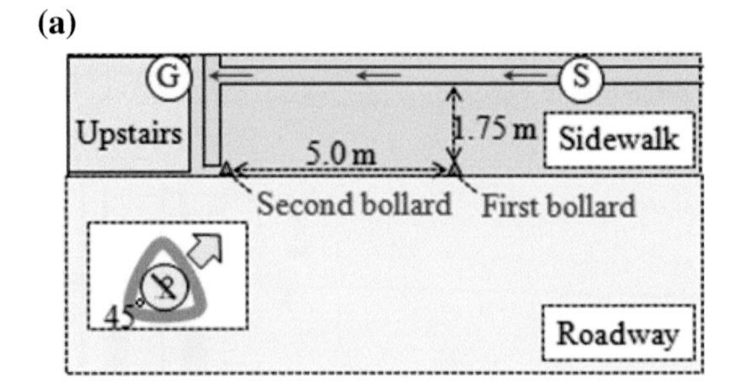

(b)

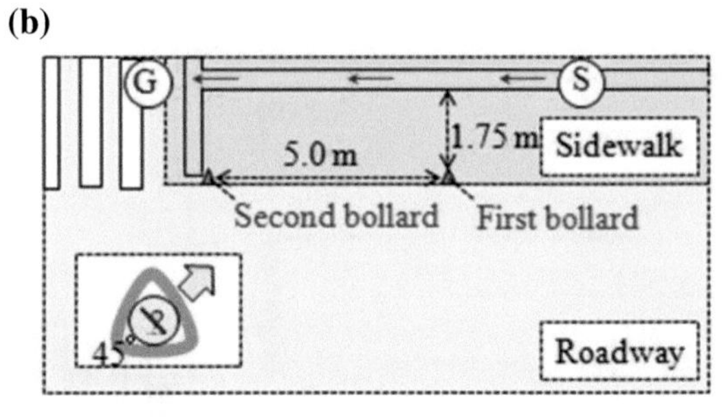

(c)

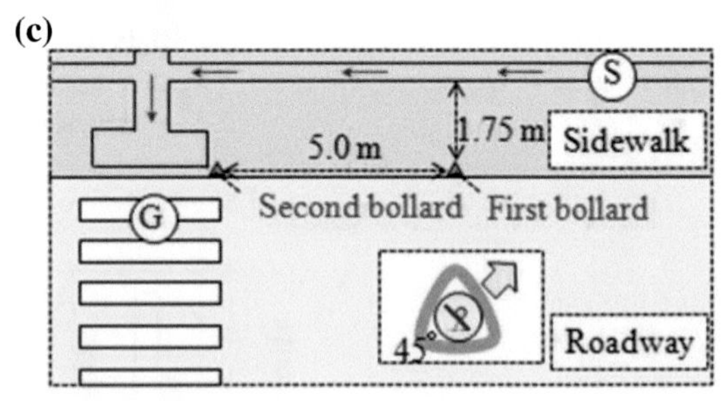

Configuration 1 is shown in Fig. 12a. The first bollard's audio message stated, "There are stairs in front of you. If you go up the stairs, there is a city hall on the right." The second bollard stated, "There are stairs in a minute."

Configuration 2 is shown in Fig. 12b. The first bollard stated, "There is a crosswalk with a signal in front of you. Its length is 20 m." The second bollard stated, "There is a crosswalk in a minute. The light is green. It will change after 15 s."

Configuration 3 is shown in Fig. 12c. The first bollard stated, "There is a crosswalk on the left about seven meters ahead. Its length is 20 m." The second bollard stated, "There is a crosswalk on the left in a minute. The light is green. It will change after 15 s."

6.2 *Experimental Results*

We interviewed the participants after they walked through the three experimental setups. Four interview questions were asked:

(1) Impression of the system: When using this system, is there a sense of security?
(2) Communication area: Can you receive the signal and perceive the light of the bollards?
(3) Audio message position and content: Are the positions and contents of the audio messages appropriate?
(4) Other concerns: How could the system be improved?

For question 1, all subjects answered that the system gave them a sense of security. Audio warnings are very effective because visually impaired people often rely on sound in daily life.

For question 2, some subjects answered that they could find the bollard in a dark road. However, they would not be able to find it if traffic in the road was heavy. Furthermore, they stated that it was difficult to walk while hold up the receiving terminal. Therefore, we conclude that is difficult for them to see the light and walk while holding up the receiving terminal.

For question 3, all subjects answered that the audio message positions and content were appropriate.

For question 4, some subjects noted that because they wanted to have their hands free, holding up the receiving terminal is not ideal. In addition, they would like to use this system at the entrance to buildings and bus stops.

In this evaluation, visually impaired people used the proposed system and evaluated its effectiveness. In addition, there were no significant differences in their answers for the three sidewalk models.

7 Conclusion

We constructed a pedestrian support system using VLC with illuminated bollards. Considering the opinions of visually impaired and elderly people collected in previous research, we proposed the pavement configuration and constructed a navigation system.

First, we experimentally evaluated the visibility of the illuminated bollards for pedestrians with low and normal vision. As a result, we determined the best illumination pattern for the bollard. Furthermore, we designed the receiving terminal to be easy to hold and manage. Moreover, we determined the communication area. From the above, we developed a prototype of the navigation system using VLC. Visually impaired people used the system and clarified its effectiveness as well as identified some areas for improvement.

References

1. Ministry of Health, Labour and Welfare. http://www.mhlw.go.jp/toukei/saikin/hw/shintai/06/ (2006). Accessed May 2014
2. The Minister of Land, Infrastructure, Transport and Tourism, Comprehensive Transport Policy. http://www.mlit.go.jp/common/000144606.pdf (2011). Accessed May 2014
3. Kii, H., Murata, Y., Oshiba, S., Nagai, Y., Watanabe, H., Iki, S., Kitani, Y., Kuwahara, N., Morimoto, K.: Accessible optical wireless pedestrian-support systems for individuals with visual impairment. In: IIAI 3rd International Conference on Advanced Applied Informatics (IIAIAAI 2014) (2014). doi:10.1109/IIAI-AAI.2014.156
4. Ohara, Y., Kado, N., Kii, H., Watanabe, H., Iki, S., Murata, Y., Nagai, Y., Anai, T., Iwamatsu, N., Kitani, Y., Kuwahara, N., Morimoto, K., Oshiba, S.: Accessible optical wireless system using LED bollard. In: 13th Forum on Information Technology (FIT 2014), no. RJ-005, pp. 33–38 (2014)
5. Itoh, M., Oshiba, S., Seki, S.: Experimental analysis of E/O and O/E conversion performances for radio-on-fiber of digital wireless communication systems. In: Asia-Pasific Microwave Conference (APMC 2014), no. FRIG-42, pp. 1282–31284 (2014)

The Image Sharpness Metric via Gaussian Mixture Modeling of the Quaternion Wavelet Transform Phase Coefficients with Applications

Yipeng Liu and Weiwei Du

Abstract In this paper, we make use of Gaussian mixture model (GMM) to describe the coefficients distribution of the quaternion wavelet transform (QWT). Derived from the parameters in GMM, the metric is proposed to find the relationship between the image blur degree and the distribution histograms of high frequencies coefficients. Also, the metric can be applied to smooth patch detection. Finally, experiments are conducted on natural images and the reasonable results indicate that the proposed metric can exhibit better performance than three common global sharpness measurements and satisfy the visual perception in the local smooth patch detection.

1 Introduction

Quaternion wavelet transform (QWT), as a novel image analysis tool, is of some superior properties compared to discrete wavelets, such as nearly shift-invariant wavelet coefficients and the ability of texture presentation which provides richer image texture information because of the phases. Bulow originally constructs the 2D analytic signal [1], and extends the concept of quaternion Fourier transform to quaternion Gabor transform (QGT) which is applied in disparity estimation and texture segmentation as the predecessor of QWT applications. Since Chan et al. [2] extended the idea of dual tree complex wavelet to quaternion domain with the application to disparity estimation using the concept of 2D Hilbert transform and analytic signal, QWT has been extensively applied to image denoising [3–5],

Y. Liu (✉)
College of Information Engineering, Zhejiang University of Technology, Hangzhou, China
e-mail: liuyipeng@zjut.edu.cn

W. Du (✉)
Department of Information Science, Kyoto Institute of Technology, Kyoto, Japan
e-mail: duweiwei@kit.ac.jp

R. Lee (ed.), *Applied Computing & Information Technology*,
Studies in Computational Intelligence 619,
DOI 10.1007/978-3-319-26396-0_13

texture classification [6–8], global and local blur degree detection [9, 10] and image fusion [11, 12].

However, to our knowledge, only several literatures focus on the coefficients modeling for QWT. Gai and Liu use hidden Markov model (HMM) [3] and Copula model [8] to exploit the relationship between magnitude and phases of QWT with application to texture classification, respectively. Liu models the magnitude as Rayleigh distribution in [4] and Yin models the coefficients with real-imaginary representation as generalized Gaussian distribution, moreover, the derived thresholds bring in the better denoising performance. The work in [9] simply use standard derivation to describe the phase changes between the clear and blurred image. Phases coefficients distribution with complex shape has not been paid enough attention until now.

As mentioned above, we purpose to explore the potential of Gaussian mixture model (GMM) inspired by the bimodal characteristics of phase coefficients. The basic idea underlying our method is that we derive the metric by means of multiple Gaussian parameters to depict the relationship between phases change and blur degree by means of analyzing the histogram of QWT coefficients of natural images. Likewise, the metric is capable of detecting relatively smooth patches in images.

The paper unfolds as follows. A brief review about QWT is given in Sect. 2. After the formulation of GMM and the proposed metric in Sect. 3, we provide the experimental validations on different applications using standard natural images and relevant discussions. Finally, Sect. 5 presents the conclusions.

2 Quaternion Wavelet Transform

For convenience of further discussions, we briefly review some basic ideas on quaternion and construction of QWT.

The quaternion algebra was invented by Hamilton in 1843 which is a generalization of the complex algebra.

$$H = \{q = a + bi + cj + dk | a, b, c, d \in \mathrm{R}\} \tag{1}$$

where the orthogonal imaginary numbers i, j and k and satisfy the following rules.

$$i^2 = j^2 = k^2 = -1, ij = k, jk = i, ki = j$$

An alternative representation for a quaternion is

$$q = |q|e^{i\phi}e^{k\psi}e^{j\theta} \tag{2}$$

where $(\phi, \psi, \theta) \in [-\pi, \pi) \times [-\pi/2, \pi/2) \times [-\pi/4, \pi/4)$. It is defined by one modulus and three angles that we call phase. The computational formula [1] is

$$\begin{cases} \phi = arctan(\frac{2(ac+bd)}{a^2+b^2+c^2+d^2}) \\ \psi = arctan(\frac{2(ab+cd)}{a^2-b^2+c^2-d^2}) \\ \theta = \frac{1}{2} arcsin(2(bc-ad)) \end{cases} \tag{3}$$

The quaternionic analytic signal is defined by its partial (H_1, H_2) and total (H_T) Hilbert transforms (HT),

$$\begin{aligned} f_A(x,y) = f(x,y) + iH_1(f(x,y)) + jH_2(f(x,y)) \\ + kH_T(f(x,y)) \end{aligned} \tag{4}$$

We start with real separable scaling function φ and mother wavelets η^H, η^V, η^D, for separable wavelet, $\eta(x,y) = \eta(x)\eta(y)$. According to the definition of the quaternionic analytic signal, the QWT, i.e. the analytic 2D wavelets can be constructed as follows.

$$\begin{cases} \varphi = \varphi_h(x)\varphi_h(y) + i\varphi_g(x)\varphi_h(y) + j\varphi_h(x)\varphi_g(y) \\ \quad + k\varphi_g(x)\varphi_g(y) \\ \eta^H = \eta_h(x)\varphi_h(y) + i\eta_g(x)\varphi_h(y) + j\eta_h(x)\varphi_g(y) \\ \quad + k\eta_g(x)\varphi_g(y) \\ \eta^V = \varphi_h(x)\eta_h(y) + i\varphi_g(x)\eta_h(y) + j\varphi_h(x)\eta_g(y) \\ \quad + k\varphi_g(x)\eta_g(y) \\ \eta^D = \eta_h(x)\eta_h(y) + i\eta_g(x)\eta_h(y) + j\eta_h(x)\eta_g(y) \\ \quad + k\eta_g(x)\eta_g(y) \end{cases} \tag{5}$$

The 2D HT's is equivalent to 1D HT's along rows and/or columns. Considering the 1D Hilbert pair of wavelets $(\eta_h, \eta_g = H\eta_h)$ and scaling function $(\varphi_h, \varphi_g = H\varphi_h)$, the analytic 2D wavelets are written in terms of separable products. Each sub-band of the QWT can be seen as the analytic signal associated with a narrow band part of the image. The QWT magnitude $|q|$, with the property of near shift-invariance, represents features at any spatial position in each frequency sub-band, and the three phases (ϕ, ψ, θ) describe the 'structure' of those features. More details about implementation of QWT used here are referenced to the work [2].

3 The Proposed Metric via Gaussian Mixture Model

Definition 1 (*Gaussian Mixture Model, GMM*) with the following probabilistic distribution,

$$P(y|\gamma) = \sum_{k=1}^{K} \alpha_k \lambda(y|\gamma_k) \tag{6}$$

where K is the quantity of Gaussian mixture components, α_k is the weight for a given mixture component $\lambda(y|\gamma_k)$, and satisfies $\alpha_k \geq 0, \sum_{k=1}^{K} \alpha_k = 1$; $\lambda(y|\gamma_k)$ is Gaussian probabilistic density function $\lambda(y|\gamma_k) = \frac{1}{\sqrt{2\pi}\sigma_k} e^{-\frac{(y-\mu_k)^2}{2\sigma_k^2}}$.
Expectation maximization (EM) algorithm is expected to estimate the probabilistic model parameters with hidden variants. For GMM, hidden variants reflect the source of the observed data y_i from the kth mixture model, noted as $\hat{\xi}_{jk}$. Firstly, initialize the model parameters, by means of the iteration, and the estimation for $\hat{\xi}_{jk}$ is

$$\hat{\xi}_{jk} = \frac{\alpha_k \lambda(y_j|\gamma_k)}{\sum_{k=1}^{K} \lambda_k \lambda(y_j|\gamma_k)} \tag{7}$$

A group of updated iterative parameters is shown in the formula (8).

$$\begin{cases} \mu_k = \frac{\sum_{j=1}^{N} \hat{\xi}_{jk} y_j}{\sum_{j=1}^{N} \hat{\xi}_{jk}} \\ \hat{\sigma}_k^2 = \frac{\sum_{j=1}^{N} \hat{\xi}_{jk} (y_j - \mu_k)^2}{\sum_{j=1}^{N} \hat{\xi}_{jk}} \\ \hat{\alpha}_k = \frac{\sum_{j=1}^{N} \hat{\xi}_{jk}}{N} \end{cases} \tag{8}$$

where N is the number of observed data. $\hat{\sigma}_k^2$ represents the width of individual Gaussian distribution, and the weight coefficient α_k indicates the corresponding proportion in the whole mixture model.

Take one image in Fig. 1a as an example, the phase ϕ is obtained in Fig. 1b after one level decomposition of QWT. It can be straightforward observed that high frequency coefficients distribution presents bimodal shape going through the non-linear computation in the formula (3). Thus, unimodal probability density functions such as Gaussian, generalized Gaussian, and Laplacian etc. are not suitable to describe it. In virtue of multiple mixture models, it is expected to estimate the complex coefficients distribution.

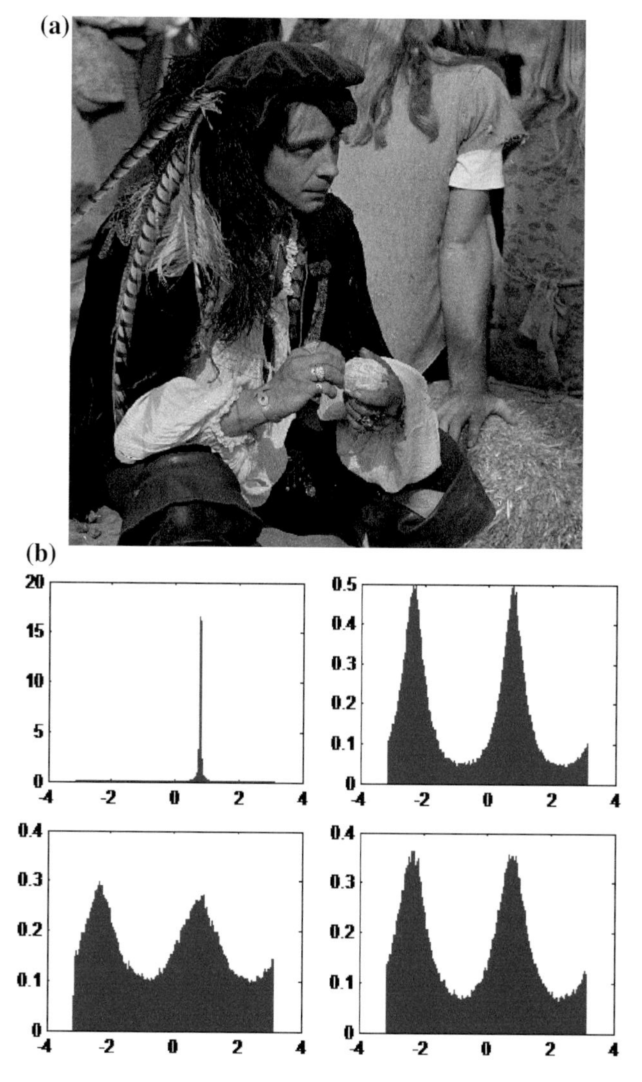

Fig. 1 **a** The test image; **b** ϕ of QWT decomposition for (**a**), and four subgraphs represent LL, HL, LH and HH wavelet subbands

The imaging systems often suffer from blur owing to defocus effect. Measuring the global or local blur degree is important to restore the blurred image. The blurred image, i.e. the sharpness of the original image is degraded, results in that the pixel intensity of texture is approach to uniformity [9]. More clearly visualized, in Fig. 2a, the Fig. 1a is processed by Gaussian blur with the window size 11×11, and the QWT HL histograms of ϕ in Fig. 2b are much narrower than those of ϕ for the

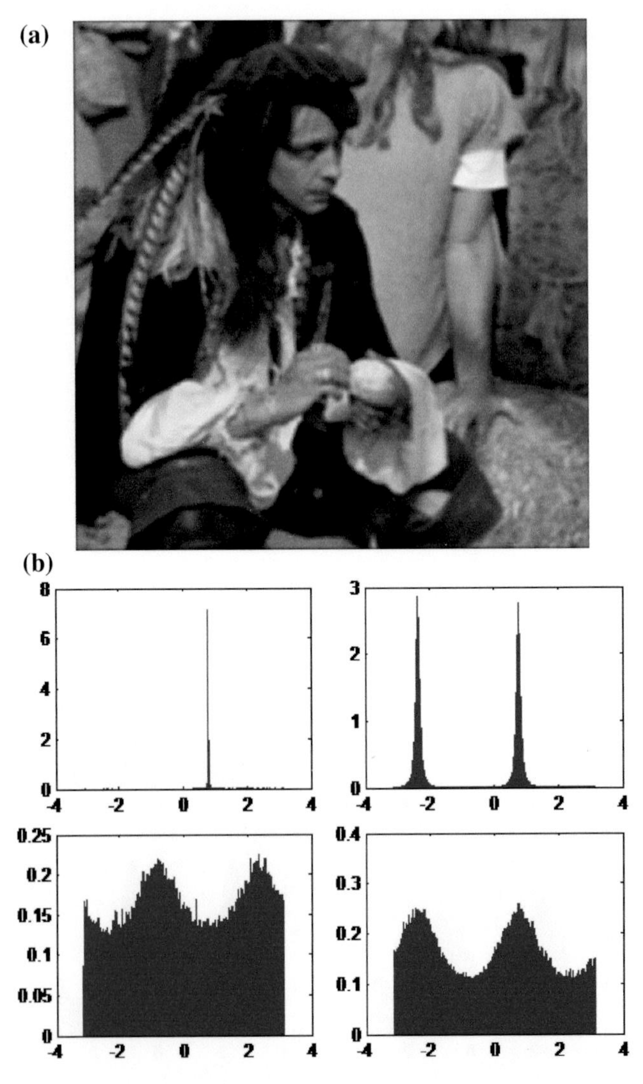

Fig. 2 **a** The blurred image; **b** ϕ of QWT decomposition of (**a**), using HL and LH to represent the *right upper* and *left below* histogram in (**b**)

original image. The probability distribution curve width of phases high frequency coefficients reflect the image blur degree, based on this, in order to detect such phase changes, the image sharpness metric is defined as

$$M_h = \sqrt{\sum_{j=1}^{K}((\hat{\alpha}_k \hat{\sigma}_k^2)_{h_1} + (\hat{\alpha}_k \hat{\sigma}_k^2)_{h_2})} \tag{9}$$

where h means high frequency, the subscript (h_1, h_2) show QWT phases HL and LH components. The parameters are derived from the Gaussian mixture model in the formula (8).

4 Experimental Results

In this section, we analyze the effect of the number of GMM and apply the proposed metric into two areas:

- Global image sharpness detection
- Local image smooth patches detection.

4.1 *The Effect of the Number of GMM*

Notice that there is one optional parameter K in the formula (6) which influences the modeling accuracy and hidden unit estimation efficiency. From the bimodal shape of phase high frequency coefficients, we firstly make $K = 2$ straightforwardly. The result for estimating GMM by EM is shown in Fig. 3. To clearly observe the relationship between GMM and its components, Fig. 4 is given. It is obvious that $K = 2$ is not a wise decision to precisely describe the phase coefficients distribution.

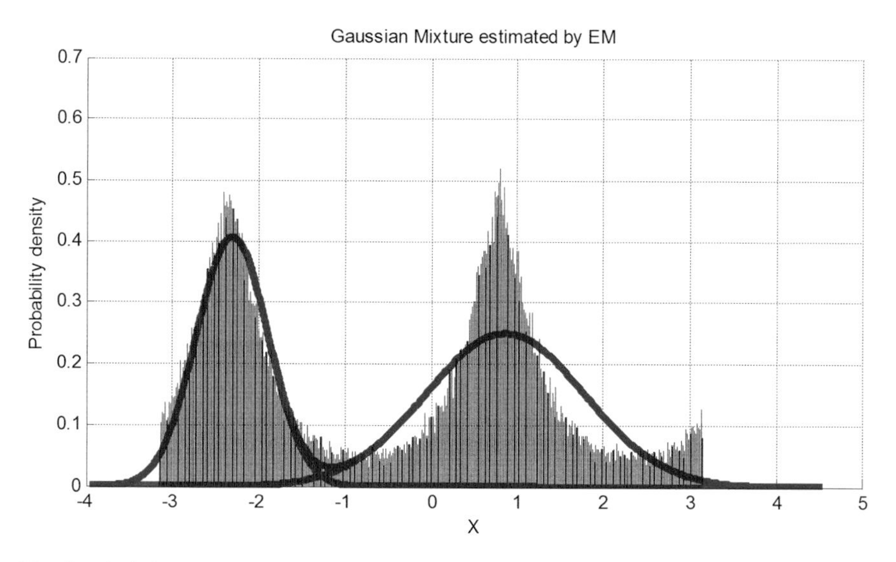

Fig. 3 *Black bars* show the phase coefficients distribution, and the *blue curve* represents the GMM result while the unconspicuous *red* one is individual Gaussian mixture component

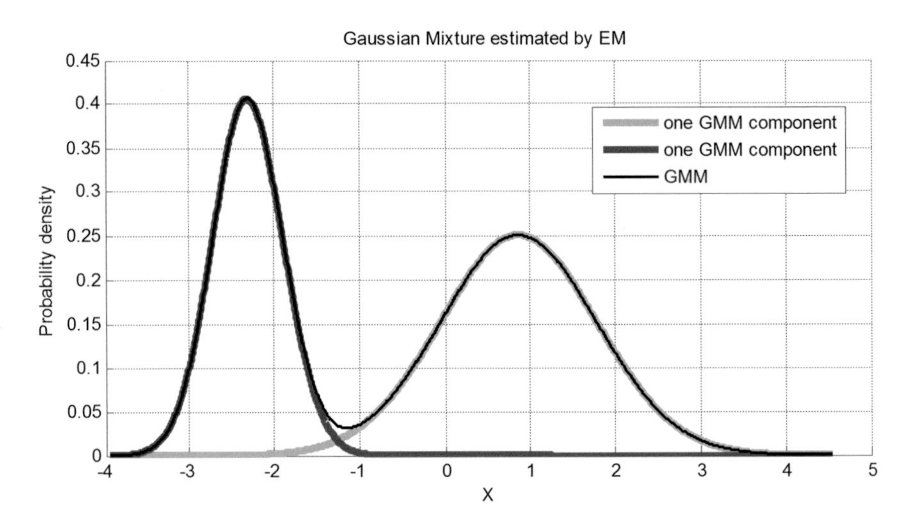

Fig. 4 Estimated GMM with two components

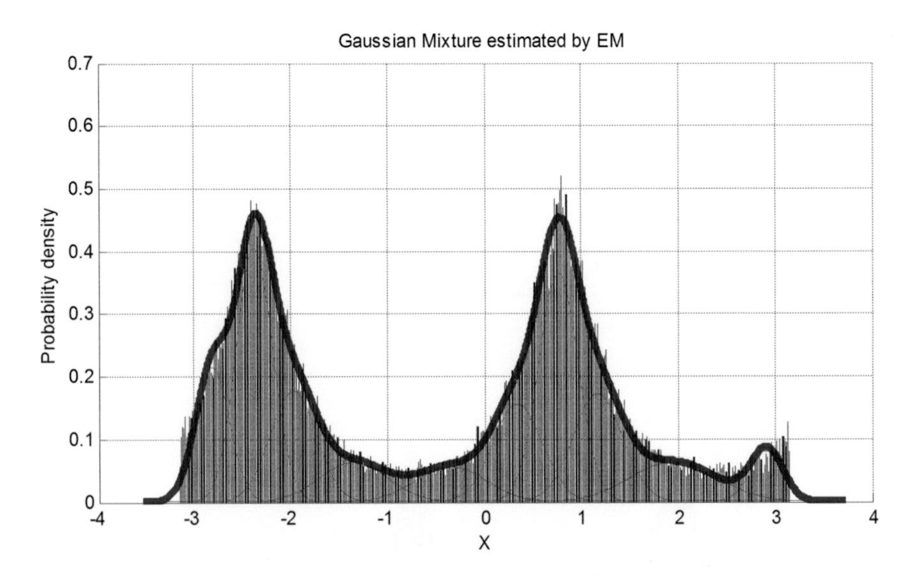

Fig. 5 GMM with 10 components. Curves captions refer to Fig. 3

There is great error for the right peak and the distribution width. With the increasing mixture components from 2 to 10, the accuracy also advances (shown in Fig. 5), but the runtime rises from 1 to 15 s. We mainly care about the distribution width, so $K = 4$ is adopted to compromise between the accuracy and efficiency in the following experiments.

4.2 Applications

In this subsection, we apply the proposed metric into following fields: one is global image sharpness detection, the other is local image smooth patches detection.

4.2.1 Global Image Sharpness Detection

The test images are generated by convolving the source images with Gaussian filter with the variable mask sizes. Here, Gaussian function plays the role of blur in the simulations with different window sizes corresponding to the blur degree.

We list following three common sharpness metrics [13] as comparative methods, and the computational results are shown in Fig. 6.

(a) L_2 norm of image gradient
This metric is also called Energy of image gradient (EOG).

$$M_1 = \sum_x \sum_y (f_x^2 + f_y^2) \tag{10}$$

(b) $L1$ norm of second derivatives of an image

$$M_2 = \sum_x \sum_y (|f_{xx}| + |f_{yy}|) \tag{11}$$

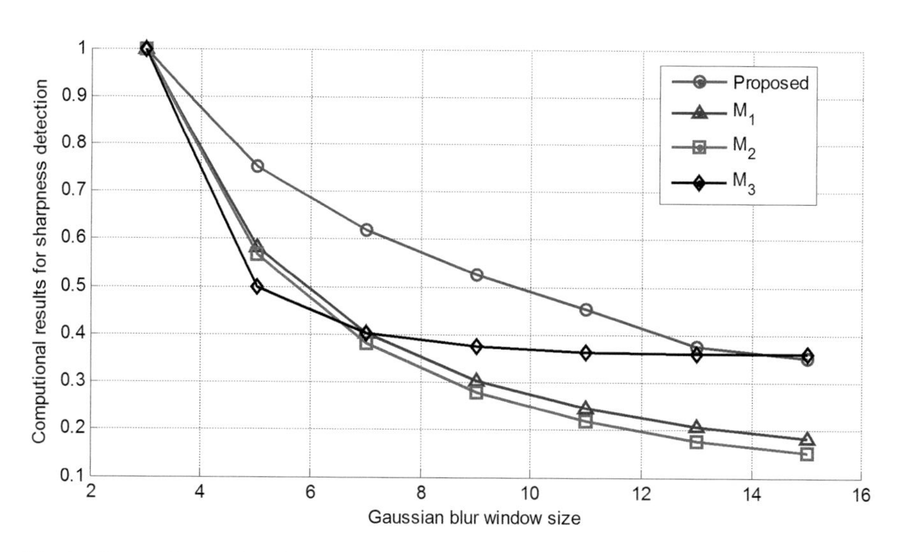

Fig. 6 Bigger window size, higher blur degree, and lower image sharpness. So the computational results monotonically decrease

(c) Energy of Laplacian of the image (EOL)

$$M_3 = \sum_x \sum_y (f_{xx} + f_{yy})^2 \quad (12)$$

For M_1, f_x and f_y denote the image gradient along the column and row directions, respectively. For M_2 and M_3, f_{xx} and f_{yy} are the second derivatives of the image.

The ideal metric curve should have the steady slope when the blur degree changes, thus the curve can differentiate images with different blur degrees. For one method, it is useful to compare the results with different blur windows size rather than different methods. The relative value between blurred image and the original image is considered. For M_3, the curve plummet to saturation so that the metric

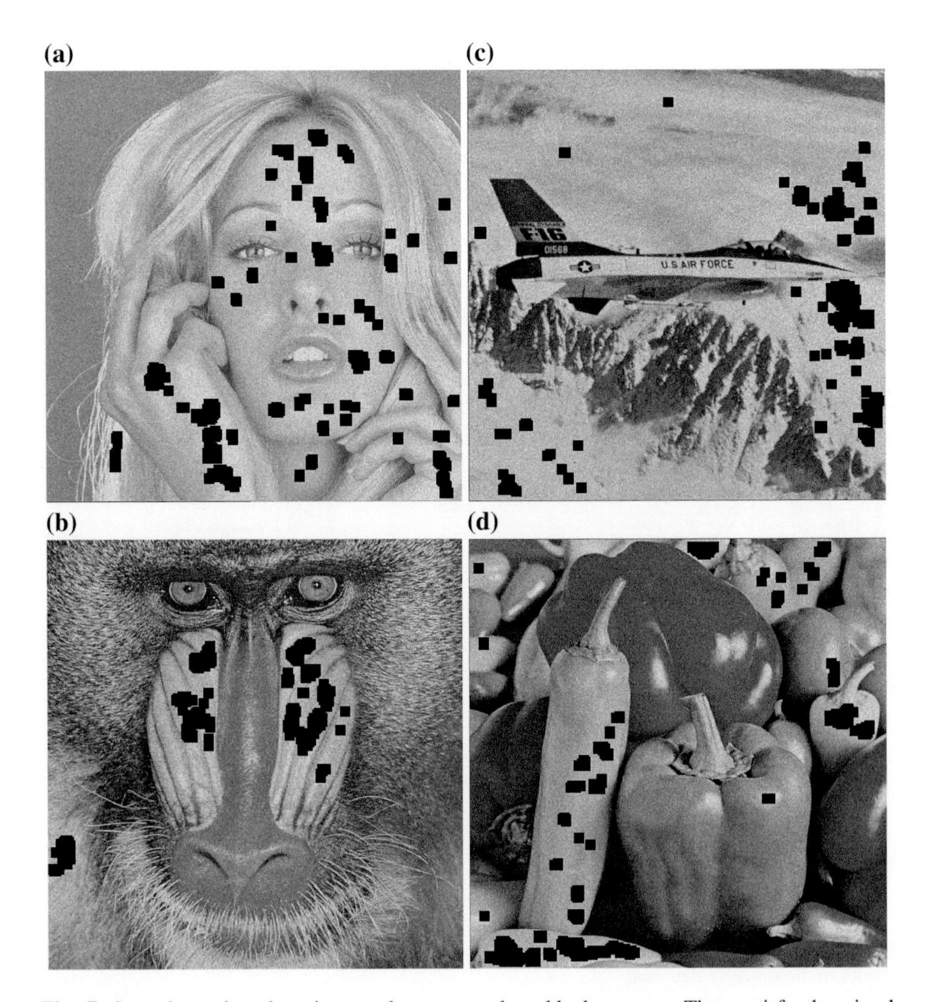

Fig. 7 Smooth patches detection results expressed as *black squares*. They satisfy the visual perception of human eyes

could not differentiate blur degrees above 5×5. From this point, M_1 and M_2 are better than M_3, but still worse than the proposed metric because for the equally spaced window size, the proposed metric presents the approximate difference of the computational results. The default QWT decomposition level is one, as shown in [9], it is expected that with the increasing of the decomposition level, the sharpness detection results will be better.

4.2.2 Local Image Smooth Patches Detection

Smooth patches detection plays the key role to noise level estimation, segmentation, object extraction, etc. Actually, smooth patches are similar to the seriously blurred ones with less textures/features/edges. We take four images as examples in Fig. 7. The detection results are promising. Note that local patch does not have enough pixels for estimating the GMM parameters, so we only use the coefficients variance with values greater than 0 to substitute the GMM results, meanwhile the displayed black squares are with relatively smaller metric values that means to relatively smooth patches. The number of exhibited black squares are 500 obtained from the lower variance value.

5 Conclusions

In this paper, we propose one image sharpness metric based on the coefficients statistics of QWT phases. The metric is derived from GMM parameters of phases coefficients of distribution to describe the blur degree. The novelty of this paper comes from introducing GMM into modeling QWT phases coefficients. The process begins with transforming the image into quaternion wavelet domain. Then the magnitude-phase representation form is computed before estimating GMM parameters for QWT phases coefficients. Finally, the image sharpness measure is calculated as the weighted sum of the individual Gaussian parameter. The experimental results indicate that the proposed metric can exhibit better performance than three listing sharpness measurements and satisfy the visual perception in the smooth patches detection.

The comparative study on different mixture models (e.g. Laplacian mixed with Gaussian model) and the further error analysis will be done in the future work.

Acknowledgements This work was financially supported by the Zhejiang Provincial Natural Science Foundation (LQ15F020009).

References

1. Bulow, T.: Hypercomplex spectral signal representations for the processing and analysis of images. Ph.D. dissertation, Christian Albrechts University, Kiel, Germany (1999)
2. Chan, W.L., Choi, H., Baraniuk, R.G.: Coherent multiscale image processing using dual-tree quaternion wavelets. IEEE Trans. Image Process. **17**(7), 1069–1082 (2008)
3. Gai, S., Yang, G., Wan, M., Wang, L.: Hidden Markov tree model of images using quaternion wavelet transform. Comput. Electr. Eng. **40**(3), 819–832 (2014)
4. Liu, Y., Jin, J., Wang, Q., Shen, Y.: Phase-preserving speckle reduction based on soft thresholding in quaternion wavelet domain. J. Electron. Imaging **21**(4), 043009 (2012)
5. Yin, M., Liu, W., Shui, J., Wu, J.: Quaternion wavelet analysis and application in image denoising. Math. Probl. Eng. 493976 (2012)
6. Soulard, R., Carre, P.: Quaternion wavelets for texture classification. Pattern Recogn. Lett. **32** (13), 1669–1678 (2011)
7. Gai, S., Yang, G., Zhang, S.: Multiscale texture classification using reduced quaternion wavelet transform. Int. J. Electron. Commun. **67**(3), 233–241 (2013)
8. Liu, C., Li, J., Fu, B.: Magnitude-phase of quaternion wavelet transform for texture representation using multilevel copula. IEEE Sig. Process. Lett. **20**(8), 799–802 (2013)
9. Liu, Y., Jin, J., Wang, Q., Shen, Y.: Phases measure of image sharpness based on quaternion wavelet. Pattern Recogn. Lett. **34**(9), 1063–1070 (2013)
10. Liu, Y., Jin, J., Wang, Q., Shen, Y., Dong, X.Q.: Region level multi-focus image fusion using quaternion wavelet and normalized cut. Signal Process. **97**(4), 9–30 (2014)
11. Xu, Y., Yang, X., Song, L., et al.: QWT: retrospective and new spplications. In: Bayro-Corrochano, E., Scheuermann, G. (eds.) Geometric Algebra Computing, pp. 249–273. Springer, London (2010)
12. Liu, Y., Jin, J., Wang, Q., Shen, Y.: Novel focus region detection method for multifocus image fusion using quaternion wavelet. J. Electron. Imaging **22**(2), 023017 (2013)
13. Huang, W., Jing, Z.: Evaluation of focus measures in multi-focus image fusion. Pattern Recogn. Lett. **28**(4), 493–500 (2007)

Generating High Brightness Video Using Intermittent Illuminations for Dark Area Surveying

Hitomi Yoshiyama, Daisuke Iwai and Kosuke Sato

Abstract When operating a remote control robot in the dark through its onboard camera, the suitable illuminations are necessary. In the situation, battery life of illuminations to survey the environment may determine the total working time. Therefore, we reduce power consumption of illuminations by switching intermittently and propose to enhance the brightness of video captured by a camera as if all illuminations are lighting constantly. In the proposed method, dark areas in a captured image are replaced with well-lighted areas in neighboring frame images by searching corresponding areas including same parts of objects. Experiments to confirm operation of the proposed method show that an image sequence as if illuminations are always lighting can be generated in the proposed method.

Keywords Tele-operation · Video image enhancement · Robot surveying

1 Introduction

Remote control robots are introduced to various works instead of human, such as in inquiry at the time of disaster, search exploration in geology, checkup of structures [1, 2, 3, 4]. In remote controlling, an operator needs to understand situation of robot and makes a judgement on what operation to do next. According to study about the situation recognition in remote control of robot, operators controlling robot remotely spent 30 % of their operating time only in situation awareness [5].

H. Yoshiyama (✉) · D. Iwai · K. Sato
Graduate School of Engineering Science, Osaka University, Machikaneyama 1-3, Toyonaka, Osaka 560-8531, Japan
e-mail: yoshiyama@sens.sys.es.osaka-u.ac.jp

D. Iwai
e-mail: daisuke.iwai@sys.es.osaka-u.ac.jp

K. Sato
e-mail: sato@sys.es.osaka-u.ac.jp

R. Lee (ed.), *Applied Computing & Information Technology*,
Studies in Computational Intelligence 619,
DOI 10.1007/978-3-319-26396-0_14

To grasp situation, visual information from a camera mounted on robot plays a major role. Operators can not recognize scene through video from a color camera in the dark such as inside without illuminations or cave where sunlight does not reach. Therefore, it is hard to operate robots in the dark. As means for obtaining visible image, use of illuminations is considered. Locating multiple illuminations enables illuminations to light widely, and then, image sequences have less shadow. At this time, it is desirable that illuminations are battery-powered illuminations because wires sometimes entangle or disturb movement of robot. However, if battery-powered illuminations are used, a problem that battery life can be limitation of time for searching activity arises.

On the other hand, an infrared camera can be used in order to obtain images in the dark. Even when using the infrared camera, powerful infrared ray with infrared illuminations is necessary, so there is also the battery problem. Although there is a method of visualizing thermal infrared ray radiated from objects without using illuminations, obtaining clear image to recognize surroundings is difficult in this case.

In this study, we light surroundings at high luminance and for a long time by lighting illuminations intermittently. Using the intermittent illuminations allows to reduce power consumption per unit time as compared with constantly lighting. Furthermore, lighting time is a moment when using intermittent illuminations, so heat dissipating of intermittent illuminations can be simplified, and it enables illuminations to light with strong impulse current at high luminance (Fig. 1). In image captured under multiple intermittent illuminations, each area that is lighted by illuminations changes intermittently. In addition, when all illuminations are extinguished, there are completely dark images. Therefore, it is difficult to recognize surroundings visually under intermittent illuminations compared with under constant illuminations.

The aim of this research is to generate high visibility image sequences by brightness compensation under intermittent illuminations. In this paper, we use the word “brightness compensation” as generating image sequences equivalent to that captured under constant illuminations from image sequence under intermittent illuminations (Fig. 2).

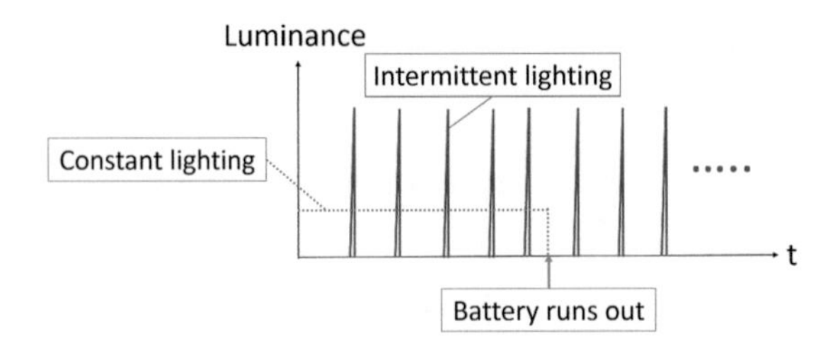

Fig. 1 Comparison of intermittent illumination with constant illumination

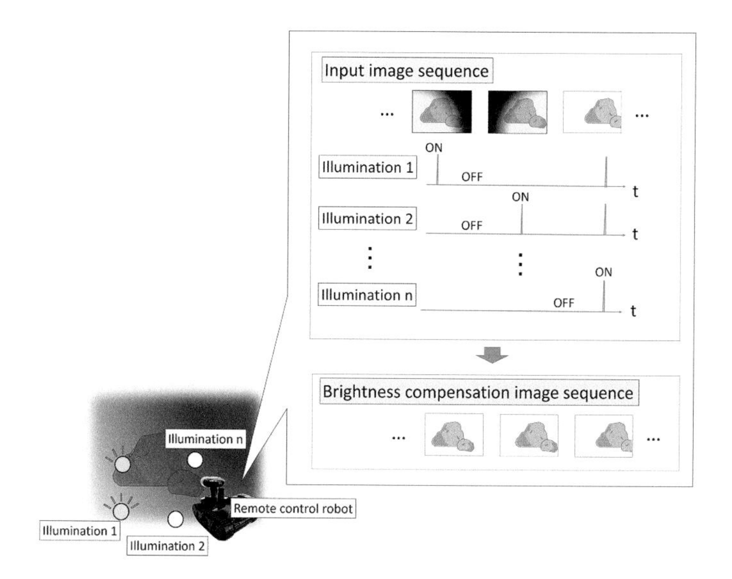

Fig. 2 Supposed situation

2 Related Research

Enhancement of dark image is often seen in image processing area. Histogram equalization is used for contrast enhancement. With histogram equalization, low contrast images are improved and become more visible [6]. Kim et al. [7] and Malm et al. [8] enhance the visibility of extremely low light images based on information of pixel values of its own image. In their method, dynamic range of each input image is increased while avoiding to expand noises. These approaches can enhance the visibility of quite low light images. However, under our intended environment using intermittent illuminations, there are images having little information or completely dark images. Therefore, the visibility of such images having little information can not be enhanced with their methods. In this study, we judge whether a captured image has any lighted areas or not, and manage completely dark images by additional processing.

Raskar et al. [9] enhance brightness of night image captured by a static camera by doing processing in the gradient domain that preserve important information such as human figure while lightening aliasing and ghosting. Li et al. [10] split important area such as highlight area or moving object area from a nighttime image, and enhance nighttime image by combining such extracted important area and daytime background image. Yamasaki et al. [11] split captured images into illumination layer and reflectance layer. Then, illumination of nighttime images are improved by using that of daytime image so as to become similar to that of daytime image. In these methods mentioned above, dark images captured at night can be enhanced to well-lighted images such as daytime images. However, these methods assume to use a static camera capturing same scene all day long. Location of objects in captured images changes because a camera is mounted on a moving robot, so such images can not be enhanced with these methods. We compose images after calculating movement of a camera.

Our approach is related to methods to generate high dynamic range (HDR) videos in replacing low-exposure area in one image with high-exposure area in adjacent frames [12, 13]. In most work on HDR, images captured with multiple exposures are combined and dynamic range of images expands. HDR methods are different from our intended environment because images dealt in HDR methods are captured in the same lighting conditions. We assume to use multiple intermittent illuminations, so lighting conditions change in every image frame. Our work is also related to methods using flash and no-flash images in terms of using light and dealing with quite different lighting images [14, 15]. However, these methods to combine flash and no-flash images are different from our approach for the intended situation. They assume existence of ambient lights and they can capture a little visible image even without flashlight, while we try to enhance image captured under environment having no or weak ambient illuminations.

3 Proposed Method of Enhancing Video Brightness

In this study, the multiple intermittent illuminations are supposed to be scattered at random locations in environment and light at arbitrary timing. Scene is captured by an RGB camera on robot. All objects in photographic range are assumed to be static, so movement of scene is only caused by movement of a camera.

In the proposed method, dark areas in a captured image are replaced by well-lighted areas in other neighboring frame images. Figure 3 shows the outline of processing of the proposed method. Image sequence captured under intermittent

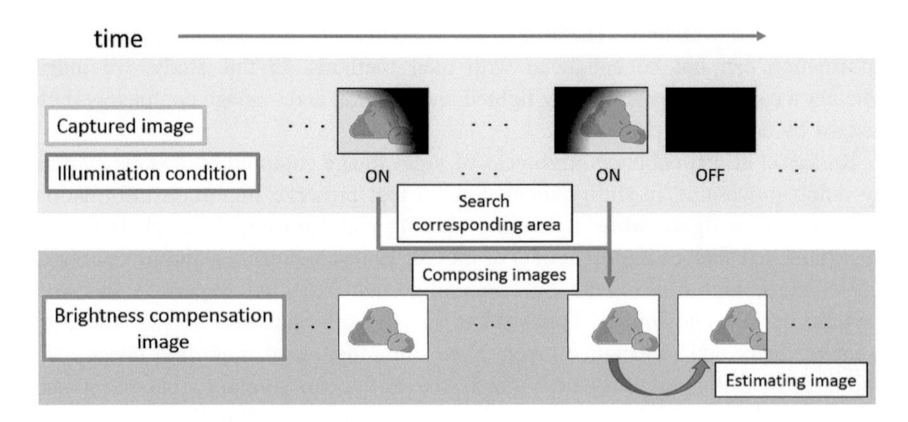

Fig. 3 Outline of image processing

illuminations consist of two types of image, *Illumination-on images* which contain well-lighted area lighted by intermittent illuminations and *Illumination-off images* which have no well-lighted area. Since we propose to enhance brightness of image by combining well-lighted areas in illumination-on images, a captured image is discriminated whether it is illumination-on image or not. We specify the algorithm for each of the following two states of a captured image.

Case1: A captured image is 'illumination-on' image. Since a camera is moving, we calculate motion vectors of objects from the last image to current captured image. Based on the motion vectors, corresponding area including the same parts of objects is estimated. Lighter corresponding area is combined into brightness compensation image.
Case2: A captured image is 'illumination-off' image. In this case, there is no well-lighted area in the image, so we estimate current brightness compensation image by using the last brightness compensation image combining illumination-on images. Based on motion vectors between last two illumination-on images, the position of each object in the current image is estimated. Using the estimated position and brightness compensation image composing last two illumination-on images, the current brightness compensation image is generated.

3.1 Discriminate Between Well-Lighted Image and Dark Image

An image sequence consists of illumination-on images and illumination-off images, so firstly we judge whether a current captured image is illumination-on image or not.

A captured image is divided into N * N blocks. We define average of brightness of ith block $(0 < i \leq N^2)\overline{l_i}$, threshold of discriminating whether a block is well-lighted or not Th_{av}. If a current image has a well-lighted ith block, that is $\overline{l_i} \geq Th_{\mathrm{av}}$, the image is defined as illumination-on image. On the other hand, if a current image has no well-lighted block, that is $\overline{l_i} < Th_{\mathrm{av}}$ for all i, the image is defined as illumination-off image.

3.2 Enhance Brightness of Well-Lighted Image

3.2.1 Calculate Corresponding Area

When a camera on the robot is moving around, the view positions are slightly different between the captured illumination-on image and the last illumination-on image. Therefore, the motion of a camera between the two images is required to consider, and after that, brightness between corresponding areas is compared. In the proposed method, corresponding areas including same parts are calculated based on motion vectors from the last illumination-on image to current captured image with block matching based on Normalized Cross Correlation (NCC). Thus, motion of a camera is estimated based on positional relation between corresponding areas. At this time, search range of corresponding area of each block is limited. The search range is area expanded to each side by r times from original block position in the coordinate system of the last illumination-on image.

3.2.2 Prevent Calculating Wrong Motion Vector

In image sequence captured by a moving camera under intermittent illuminations, calculating correct motion vectors is not stable due to change of lighting condition and position of a camera. When motion vector of a certain block is unreliable, motion vector of that block is estimated based on motion vectors of neighboring blocks under assumption that motion vectors of neighboring blocks are similar to each other.

The way to judge whether motion vector of a certain block is reliable or not is as follows. First, when the average luminance of a block is smaller than threshold Th_{av}, motion vector of that block is considered unreliable because a pixel value distribution used for block matching is possible to be lost. Also if the degree of similarity between attention block and corresponding area in block matching is low, that motion vector of attention block is considered unreliable. Specifically, if value of NCC in block matching between attention block and corresponding area in the last illumination-on image is more than threshold Th_{NCC}, motion vector of attention block is considered reliable. On the other hand, value of NCC in block matching between attention block and corresponding area in the last illumination-on image is smaller than threshold Th_{NCC}, motion vector of attention block is considered unreliable.

If motion vector of a block is judged to be unreliable, motion vector of attention block is estimated based on motion vectors of neighboring blocks. Motion vector of attention block is replaced by that of nearest block whose matching result is reliable. In the case that there are more than one nearest block whose matching result is reliable, motion vector of attention block is replaced by motion vector of nearest block that has maximum similarity. If matching results of all blocks are unreliable, motion vectors of all blocks are replaced by motion vector of block whose value of NCC is maximum.

3.2.3 Enhance Brightness of Image

Based on motion vectors between two illumination-on images, the average luminance of attention block is compared with the average luminance of corresponding area in the last illumination-on image. Pixel values in area whose average brightness is the larger is used as pixel values of block in brightness compensation image.

3.3 Enhance Brightness of Dark Image

If a captured image is illumination-off image, because the image does not include lighted area, it is not possible to calculate motion vector with block matching. In such a case, motion vector of each block per frame is estimated by dividing motion vector of each block between the last two illumination-on images by the number of frames existing between the last two illumination-on images. Based on estimated motion vector, position of corresponding area of each block in current illumination-off image is estimated in image made by composing the last two illumination-on images. The brightness of current illumination-off image is arranged by replacing blocks in illumination-off image with corresponding area in brightness enhanced image composing the last two illumination-on images.

4 Experiment to Test Operation

Experiments to confirm operation of the proposed method had been carried out. In this experiment, we checked whether or not image sequence similar to image sequence such as all illuminations are always lighting is possible to be generated from image sequence under intermittent illuminations with the proposed method. We call image sequence such as all illuminations are always lighting *target image sequence*. Since illumination lighting pattern has an influence on results of brightness compensation with the proposed method, illumination lighting pattern is changed as a parameter in this experiment. We used real image sequences captured under different illumination lighting patterns and evaluated results. Evaluation methods were qualitative assessment comparing brightness compensation image sequence with target image sequence by looking and quantitative assessment using peak signal-to-noise ratio (PSNR) and structural similarity (SSIM [16]) to target image sequence.

4.1 Experimental Settings

We generated the image sequences such as captured under two intermittent illuminations in the dark room for experiments.

Two illuminations (A, B), an RGB camera and two objects (A, B) were located as shown in Fig. 4 in the darkroom which had no other light source than the two intermittent illuminations. Switching on/off of the two intermittent illuminations (A, B), the scene was captured by the static RGB camera (Point Gray Research Chamereon13S2C, 1280 * 960 pixels). Repeating capturing the scene moving the position of the camera towards the right side of the objects by 2 mm/frame, total 15 cm, we generated image sequences such as were captured by the camera (30 fps) mounted on the robot at 6 cm per second under intermittent illuminations. At each

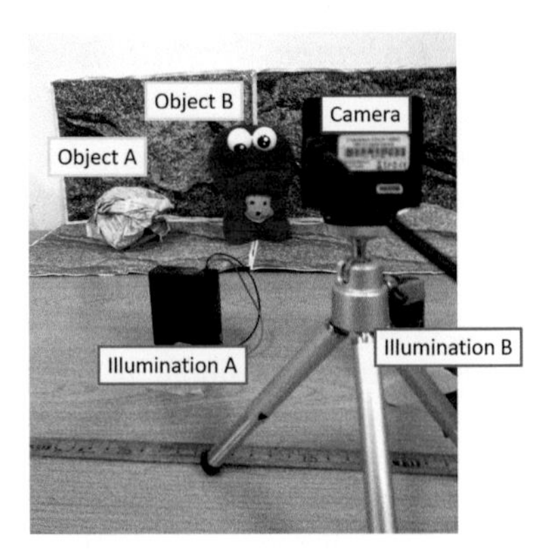

Fig. 4 Experiment envrionment

position of the camera, images whose lighting conditions were different from each other were captured. That is to say, images of three types of lighting condition: (a) both illumination A and illumination B are on, (b) the illumination A is on and the illumination B is off, (c) the illumination A is off and the illumination B is on were captured. Four images adding an image whose brightness value of all pixels was zero to the set of three images were considered as image set of each camera position. We generated input image sequences which had desired lighting conditions and all other photographing conditions were unified by choosing one image from the image set at each camera position.

Power consumption of illuminations for each input image sequence to target image was calculated by Eq. 1 based on the number of illumination-on images out of the number of all frame; n_{A} denotes the number of the images when the illumination A is on, n_{B} denotes the number of images when the illumination B is on, and n_{all} is the number of all frames.

$$(PowerConsumption) = \frac{n_{\mathrm{A}} + n_{\mathrm{B}}}{2n_{\mathrm{all}}} * 100 \tag{1}$$

In the experiment, we used eight types of input image sequence which had different power consumption. The patterns of power consumption of input image sequences are shown on Table 1. In Table 1, the lighting patterns of two

Table 1 Lighting pattern of input image sequences and power consumption

Illumination A	Illumination B	Power consumption (%)
(1, 2)	(2, 2)	50.0
(1, 2)	(2, 3)	41.4
(1, 3)	(2, 5)	26.3
(1, 5)	(2, 7)	16.4
(1, 7)	(2, 11)	10.5
(1, 11)	(2, 13)	7.23
(1, 17)	(2, 19)	5.26
Non-lighting	Non-lighting	0

illuminations are shown as "(frame number of first lighting, fixed interval of lighting)". For example, (1, 2) shows an example that the illumination lights at the 1st frame and it lights every two frames.

4.2 Processing to Enhance Brightness

We enhanced brightness of the eight types of input image sequence shown in Table 1 with the proposed method. We used a laptop PC (CPU: Intel Core i7 1.80 GHz, RAM: 4.00 GB) for processing. In the experiment, each parameter in the proposed method was experimentally determined as shown in Table 2. For speed up of processing, resolution of each image in the input image sequences was converted into 640 * 480 pixels (24 bit RGB). Processing time per frame was an average of 0.120 s for illumination-on image and an average of 0.00870 s for illumination-off image.

4.3 Result and Discussion

4.3.1 Qualitative Assessment

Figure 5 shows a result of enhancing brightness of input image sequence whose power consumption was 10.5 %. We compared the target image sequence with brightness compensation image sequences, and confirmed that image sequences whose power consumption was higher than 10.5 % were similar to target image sequence. On the other hand, in image sequences whose power consumption was lower than 7.23 %, image sequences had different position of objects from real position or significant misalignment shown in Fig. 6.

The cause of misalignment is low accuracy of matching corresponding area in image sequences captured under intermittent illuminations. The smaller power consumption becomes, the more illumination-off images exist between illumination-on images, so difference of view point location becomes bigger. Therefore, increase of error correspondence influenced the result and made conspicuous misalignment.

Another point to consider is that we estimated motion vectors in illumination-off images based on the assumption that the movement of a camera was constant. However, in real environment, the assumption is not realistic and it is possible that

Table 2 Parameter settings

Parameter	Setting
Search range r (times)	1.8
Threshold of average brightness Th_{av}	15
Threshold of NCC Th_{NCC}	0.98
The number of blocks N	8

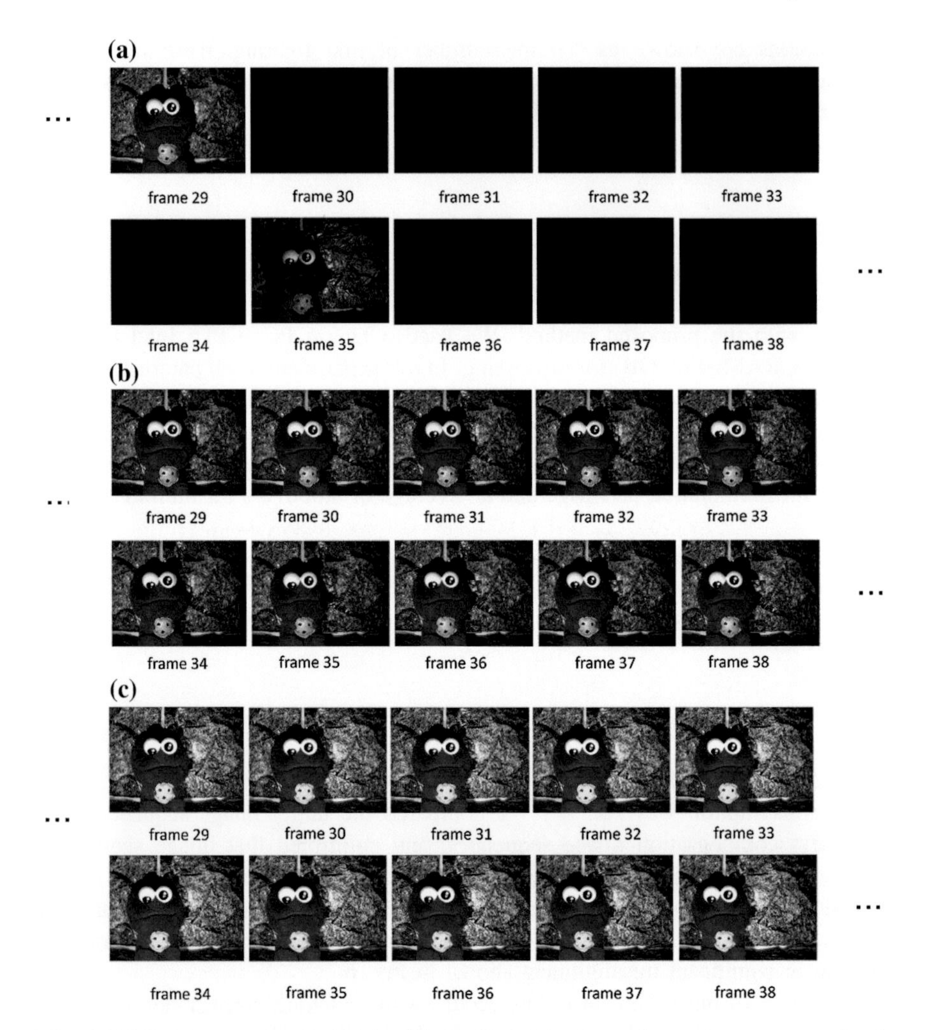

Fig. 5 Brightness compensation result of input image sequence whose power consumption is 10.5 %. **a** Input image sequence. **b** Brightness compensation image sequence. **c** Target image sequence

Fig. 6 Misalignment in brightness compensation image

actual movement of a camera is greatly different from estimated movement in the illumination-off frames. Therefore, it is necessary to consider illumination lighting pattern which was determined experimentally. For example, putting on illuminations before the difference between estimated and actual movement of a camera becomes remarkable seems to be one solution. In addition, movement of a camera in this experiment was assumed to be toward the only right side of front, but movement of a camera is generally toward the front. On this point, it is possible to apply the proposed method to forward movement by increasing precision of matching and make size of blocks smaller because forward movement is also regarded to be translational movement.

4.3.2 Quantitative Assessment

Results of enhancing brightness of image sequences shown in Table 1 were evaluated using PSNR and SSIM. PSNR and SSIM are evaluation indices of image, and PSNR indicates degree of deterioration with signal level. SSIM has correlational relationship and evaluates structural similarity. SSIM is an index for gray scale image, so each image is converted into gray scale for calculating SSIM. PSNR and SSIM of each image in input image sequence to target image sequence and PSNR and SSIM of each image in brightness compensation image sequence to target image sequence were calculated. We compared average PSNR and SSIM of each frame in image sequences. PSNR was calculated after removing same images with images of input image sequences. Figure 7 shows relationship of PSNR, and Fig. 8 shows relationship of SSIM.

It has been shown that PSNR and SSIM of brightness compensation image sequences are larger than those of input image sequences. However, PSNR and SSIM of image sequences whose power consumption is under 7.23 % are also improved, though those image sequences have remarkable misalignment. In this case, it is natural that PSNR and SSIM of input image sequences drastically decrease because input image sequences have images whose pixel values are all zero. Due to these reasons, evaluating with PSNR and SSIM is not always proper.

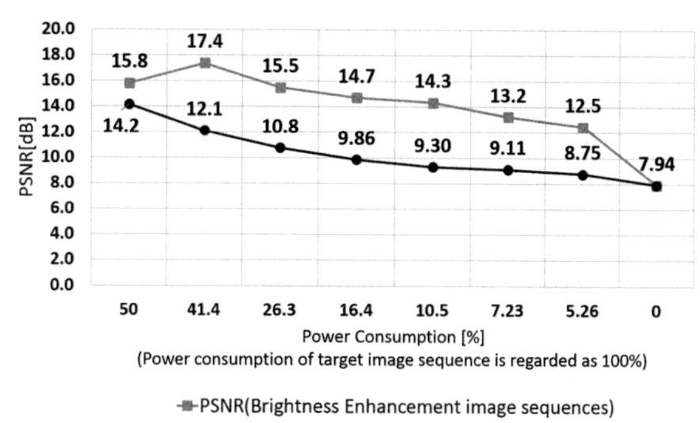

Fig. 7 Relationship between power consumption and PSNR

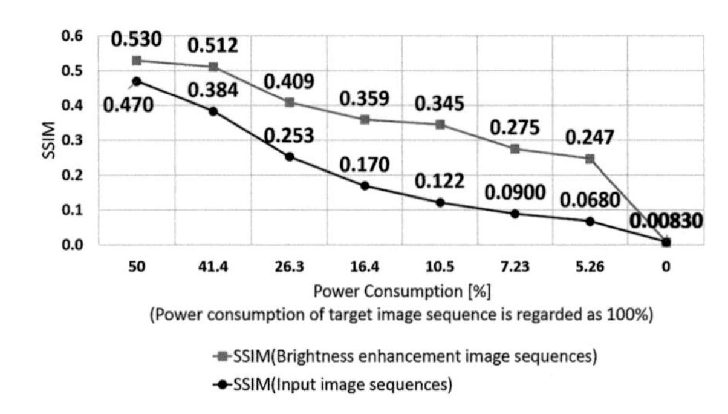

Fig. 8 Relationship between power consumption and SSIM

In the future it is desirable to evaluate by subjects because a human operator controls a robot through video, so we need user evaluation experiment.

5 Conclusion

In this study, we tried to reduce power consumption of illumination by lighting illuminations intermittently, and propose a method to enhance brightness of image sequence captured under intermittent illuminations. In the proposed method, we replace dark area in a captured image with corresponding well-lighted area in other neighboring frame image and generate image sequences such as illuminations are always lighting. By experiment to confirm operation of the proposed method, under experiment environment we validated by sight the practicability of producing similar brightness compensation image sequences to image sequences such as illuminations are always lighting.

This research is considered to be applied to various levels of surveying. If we need only rough information such as observing the structure, small amount of intermittent illuminations are necessary. On the other hand, methods to increase illuminations to a certain degree or to synchronize intermittent illuminations are considered to help us obtain detailed information. Our future plan is to improve the accuracy of extracting corresponding areas and apply the proposed method to the actual environment such as more complicated motion of a robot in various illuminations arrangement.

References

1. Kanegawa, T., Yamasaki, T., Igarashi, H., Matsuno, F.: Development of the snake-like rescue robot "kohga". In Proceedings of IEEE International Conference on Robotics and Automation, pp. 5081–5086 (2004)
2. Muscato, G., Bonaccorso, F., Cantelli, L., Longo, D., Melita, C.D.: Volcanic environments: robots for exploration and measurement. IEEE Robot. Autom. Mag. **19**(1), 40–49 (2012)

3. Nagatani, K., Kiribayashi, S., Okada, Y., Yoshida, K., Tadokoro, S., Nishimura, T., Yoshida, T., Koyanagi, E.: Emergency response to the nuclear accident at the Fukushima Daiichi nuclear power plants using mobile rescue robots. J. Field Robot. **30**(1), 44–63 (2013)
4. Zhuang, F., Zupan, C., Chao, Z., Yanzheng, Z.: A cable-tunnel inspecting robot for dangerous environment. Int. J. Adv. Robot. Syst. **5**(3), 243–248 (2008)
5. Yanco, H.A., Drury, J.: "Where Am I?" Acquiring situation awareness using a remote robot platform. In Proceedings of the IEEE Conference on Systems, Man and Cybernetics, pp. 2835–2840 (2004)
6. Abdullah-Al-Wadud, M., Kabir, M., Dewan, M., Chae, O.: A dynamic histogram equalization for image contrast enhancement. IEEE Trans. Consum. Electron. **53**(2), 593–600 (2007)
7. Kim, M., Park, D., Han, D., Ko, H.: A novel approach for denoising and enhancement of extremely low-light video. IEEE Trans. Consum. Electron. **61**(1), 72–80 (2015)
8. Malm, H., Oskarsson, M., Warrant, E., Clarberg, P., Hasselgren, J., Lejdfors, C.: Adaptive enhancement and noise reduction in very low light-level video. In Proceedings of IEEE 11th International Conference on Computer vision, pp. 1–8 (2007)
9. Raskar, R., Ilie, A., Yu, J.: Image fusion for context enhancement and video surrealism. In Proceedings of the 3rd Symposium on Non-Photorealistic Animation and Rendering, pp. 85–95 (2004)
10. Li, J., Li, S.Z., Pan, Q., Yang, T.: Illumination and motion-based video enhancement for night surveillance. In Proceedings of the 2nd Joint IEEE International Workshop on Visual Surveillance and Performance Evaluation of Tracking and Surveillance, pp. 169–175 (2005)
11. Yamasaki, A., Takauji, H., Kaneko, S., Kanade, T.: Denighting: enhancement of nighttime images for a surveillance camera. In Proceedings of 19th International Conference on Pattern Recognition, pp. 1–4 (2008)
12. Kalantari, N.K., Shechtman, E., Barnes, C., Darabi, S., Goldman, D.B., Sen, P.: Patch-based high dynamic range video. ACM Trans. Graph. **32**(6), 202 (2013)
13. Kang, S.B., Uyttendaele, M., WInder, S., Szeliski, R.: High dynamic range video. ACM Trans. Graph. **22**(3), 319–325 (2003)
14. Eisemann, E., Durand, F.: Flash photography enhancement via intrinsic relighting. ACM Trans. Graph. **23**(3), 673–678 (2004)
15. Petschnigg, G., Szeliski, R., Agrawala, M., Cohen, M., Hoppe, H.: Digital photography with flash and no-flash image pairs. ACM Trans. Graph. **23**(3), 664–672 (2004)
16. Wang, Z., Bovik, A.C., Sheikh, H.R., Simoncelli, E.P.: Image quality assessment: from error visibility to structural similarity. IEEE Trans. Image Process. **13**(4), 600–612 (2004)

Author Index

R. Lee (ed.), *Applied Computing & Information Technology*,
Studies in Computational Intelligence 619,
DOI 10.1007/978-3-319-26396-0

Printed in the United States
By Bookmasters